An Impulse and Earthquake Energy Balance Approach in Nonlinear Structural Dynamics

T0174641

An Impulse and Earthquake Energy Balance Approach in Nonlinear Structural Dynamics

Izuru Takewaki and Kotaro Kojima

CRC Press
Taylor & Francis Group
Boca Raton London New York

CRC Press is an imprint of the
Taylor & Francis Group, an **informa** business

First edition published 2021
by CRC Press
6000 Broken Sound Parkway NW, Suite 300, Boca Raton, FL 33487-2742

and by CRC Press
2 Park Square, Milton Park, Abingdon, Oxon, OX14 4RN

First issued in paperback 2022

Library of Congress Cataloging-in-Publication Data
Names: Takewaki, Izuru, author. | Kojima, Kotaro, author.
Title: An impulse and earthquake energy balance approach in nonlinear structural dynamics / Izuru Takewaki, Kotaro Kojima.
Description: First edition. | Boca Raton, FL : CRC Press, 2021. | Includes bibliographical references and index.
Identifiers: LCCN 2021000740 (print) | LCCN 2021000741 (ebook) | ISBN 9780367681401 (hardcover) | ISBN 9780367681418 (paperback) | ISBN 9781003134435 (ebook)
Subjects: LCSH: Earthquake engineering. | Force and energy.
Classification: LCC TA654.6 .T3424 2021 (print) | LCC TA654.6 (ebook) | DDC 624.1/762--dc23
LC record available at https://lccn.loc.gov/2021000740
LC ebook record available at https://lccn.loc.gov/2021000741

ISBN: 978-0-367-68141-8 (pbk)
ISBN: 978-0-367-68140-1 (hbk)
ISBN: 978-1-003-13443-5 (ebk)

DOI: 10.1201/9781003134435

Typeset in Sabon
by SPi Global, India

Contents

Preface

On April 14 and 16, 2016, two consecutive large earthquakes occurred in Kumamoto, Japan. In most countries, it is a common understanding that building structures are designed to resist once to the ground motion from a large earthquake, e.g., one with the intensity level 7 (the highest level on the Japan Meteorological Agency scale; approximately X–XII on the Mercalli scale), during its service life. However, in case of the Kumamoto earthquakes, a number of buildings were subjected to such large ground motions twice in a few days. This phenomenon was unpredictable, and the authors were convinced during and immediately after the earthquake that the critical excitation method is absolutely necessary for enhancing the earthquake resilience of building structures and engineering systems.

The senior author believed for a long time that near-fault ground motions have peculiar characteristics with a few simple waves (see Figure 0.1), and the response of buildings to such ground motions can be characterized by the response to such a simple wavelet. Furthermore, the response to such a simple wavelet can be substituted by the response to the equivalent impulse set (see Figure 0.2). The response to impulses can be described by the continuation of free vibration components, and this fact leads to the straightforward derivation of responses of even elastic-plastic structures.

In this monograph, the critical excitation problems for elastic-plastic structures under double and triple impulses are explained with the interval of impulses as a variable parameter (Kojima and Takewaki 2015a, b). Furthermore, the critical excitation problems for elastic-plastic structures under multiple impulses as a representative of long-period and long-duration ground motions are tackled with the interval of impulses as a variable parameter (Kojima and Takewaki 2015c, Kojima and Takewaki 2017). This approach can overcome the difficulty, called the nonlinear resonance, encountered first around 1960 in the field of nonlinear vibration, and the

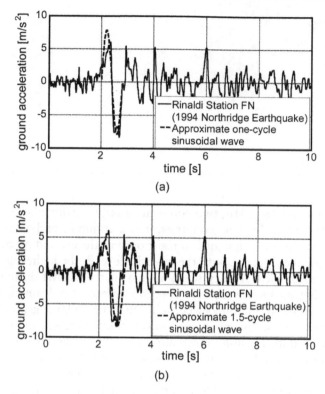

Figure 0.1 Simple modeling of Rinaldi Station fault-normal component
(Northridge EQ. 1994) as representative of near-fault ground motion:
(a) one-cycle sinusoidal wave modeling, (b) 1.5-cycle sinusoidal wave
modeling.

critical excitation problems for elastic-plastic structures are tackled in a
more direct way than the conventional methods including laborious computation (see Table 0.1). It can be said that the approach explained newly in
this monograph opened the door for an innovative field of nonlinear
dynamics.

  In principle, the method explained in this monograph is based on the
energy balance law, which is taught as part of a high school physics course.
Therefore, undergraduate students can read and understand this work. The
authors hope that this monograph is also useful for graduate students for
research and structural designers/engineers for practice.

Izuru Takewaki
Kotaro Kojima
Kyoto, 2020

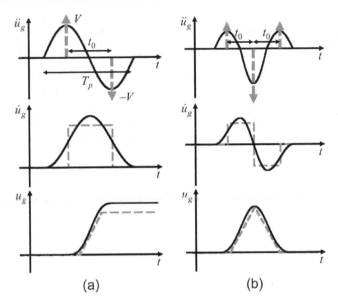

Figure 0.2 Impulse modeling of near-fault ground motion: (a) one-cycle sinusoidal wave and double impulse, (b) 1.5-cycle sinusoidal wave and triple impulse (Kojima and Takewaki 2015a).

Table 0.1 Conventional method and proposed method.

Conventional method (1960s Caughey, Iwan)	Proposed method (2015 Kojima and Takewaki)
① Steady-state	① Transient and steady-state
② Difficulty in elastic-perfectly plastic model	② Possible even for elastic-perfectly plastic (any model)
③ Inevitable repetition (equivalent parameters/resonant frequency)	③ No repetition required

<Proposed method enables>
→ Closed-form **critical** response of elastic-plastic structure
→ Derive **resonant frequency** (impulse interval) without repetition
→ Closed-form **noncritical** response of elastic-plastic structure based on closed-form **critical** response

REFERENCES

K. Kojima and I. Takewaki (2015a). Critical earthquake response of elastic-plastic structures under near-fault ground motions (Part 1: Fling-step input), *Frontiers in Built Environment* (Specialty Section: Earthquake Engineering), Volume 1, Article 12.

K. Kojima and I. Takewaki (2015b). Critical earthquake response of elastic-plastic structures under near-fault ground motions (Part 2: Forward-directivity input), *Frontiers in Built Environment* (Specialty Section: Earthquake Engineering), Volume 1, Article 13.

K. Kojima and I. Takewaki (2015c). Critical input and response of elastic-plastic structures under long-duration earthquake ground motions, *Frontiers in Built Environment* (Specialty Section: Earthquake Engineering), Volume 1, Article 15.

K. Kojima and I. Takewaki (2017). Critical steady-state response of SDOF bilinear hysteretic system under multi impulse as substitute of long-duration ground motions, *Frontiers in Built Environment* (Specialty Section: Earthquake Engineering), Volume 3, Article 41.

Authors

Izuru Takewaki is a professor of building structures in Kyoto University and president of Architectural Institute of Japan (AIJ). He also serves as the field chief editor of *Frontiers in Built Environment* and has published over 200 international journal papers. He has been awarded numerous prizes, including the Research Prize of AIJ (2004), the 2008 Paper of the Year in *Journal of the Structural Design of Tall and Special Buildings*, and the Prize of AIJ for Book (2014).

Kotaro Kojima is an assistant professor of building structures in Kyoto Institute of Technology since 2018. He obtained a PhD from Kyoto University in 2018 on the theme of Critical Earthquake Excitation Method for Elastic-plastic Building Structures using Impulse Sequence and Energy Balance Law.

Chapter 1

Introduction

1.1 MOTIVATION OF THE PROPOSED APPROACH

1.1.1 Simplification of near-fault pulse-type ground motion

The recording and documentation of earthquake ground motions started in the middle of the 20th century (PEER Center 2013). A classification of earthquake ground motions has been conducted (Abrahamson et al. 1998, Takewaki 1998, Bozorgnia and Campbell 2004), e.g., rock records, soil records, etc. Other classifications exist, namely (1) the near-fault ground motion and (2) the long-period, long-duration ground motion. While the former is well known and has been investigated for a long time, the latter is getting attention recently. In this book, both types are discussed.

Since the time typical earthquakes began to be recorded, the effects of near-fault ground motions on the structural responses have been investigated extensively (Bertero et al. 1978, Hall et al. 1995, Sasani and Bertero 2000, PEER Center et al. 2000, Mavroeidis et al. 2004, Alavi and Krawinkler 2004, Kalkan and Kunnath 2006, 2007, Xu et al. 2007, Rupakhety and Sigbjörnsson 2011, Yamamoto et al. 2011, Vafaei and Eskandari 2015, Khaloo et al. 2015, Kojima and Takewaki 2015). The concepts of fling-step and forward-directivity discussed in the field of engineering seismology are widely used to characterize such near-fault ground motions (Mavroeidis and Papageorgiou 2003, Bray and Rodriguez-Marek 2004, Kalkan and Kunnath 2006, Mukhopadhyay and Gupta 2013a, b, Zhai et al. 2013, Hayden et al. 2014, Yang and Zhou 2014, Kojima and Takewaki 2015). In particular, the San Fernando earthquake in 1971, the Northridge earthquake in 1994, the Hyogoken-Nanbu (Kobe) earthquake in 1995, and the Chi-Chi (Taiwan) earthquake in 1999 drew special attention to many earthquake structural engineers.

The fling-step (fault-parallel) and forward-directivity (fault-normal) inputs have been characterized by two or three wavelets. For this class of ground motions many useful researches have been conducted. Mavroeidis and Papageorgiou (2003) investigated the characteristics of this class of ground motions in detail and proposed some wavelets (for example, Gabor wavelet and Berlage wavelet). Xu et al. (2007) used a kind of Berlage wavelet and applied it to the evaluation of performances of passive energy dissipation systems. Takewaki and Tsujimoto (2011) employed the Xu's approach and proposed a method for scaling ground motions from the viewpoints of story drift and input energy demand. Takewaki et al. (2012) made use of a sinusoidal wave for simulating pulse-type waves. Kojima and Takewaki (2015) started a new approach using the double impulse (Kojima et al. 2015a) by picking up the principal properties of those ground motions. They focused on the intrinsic response characteristics by the near-fault ground motion (Kojima and Takewaki 2015).

Most of the previous works on the near-fault ground motions deal with the elastic response, because the number of parameters (e.g., duration and amplitude of a pulse, ratio of pulse frequency to structure natural frequency, change of equivalent natural frequency for the increased input level) to be considered is huge, and even the computation of elastic-plastic response is quite complicated.

To tackle such important but complicated problem, the double impulse input was introduced as a substitute of the fling-step near-fault ground motion, and a closed-form solution was derived of the elastic-plastic response of a structure by the "critical double impulse input" (Kojima and Takewaki 2015). It was shown that, since only the free-vibration is induced under such double impulse input, the energy balance approach plays an important role in the derivation of the closed-form solution of a complicated elastic-plastic response. It is also shown that the maximum inelastic deformation can occur either after the first impulse or after the second impulse depending on the input level of the double impulse. The validity and accuracy of the theory were investigated through the comparison with the response analysis result to the corresponding equivalent one-cycle sinusoidal input as a representative of the fling-step near-fault ground motion. The amplitude of the double impulse was modulated so that its maximum Fourier amplitude coincided with that of the corresponding one-cycle sinusoidal input. The validity and accuracy of the theory were also checked through the comparison with the elastic-plastic responses under the actual recorded near-fault ground motions.

1.1.2 Resonant response in nonlinear structural dynamics and earthquake-resistant design

The closed-form or nearly closed-form expressions for the elastic-plastic earthquake response have been obtained so far only for the steady-state response to sinusoidal input or the transient response to an extremely simple sinusoidal input (Caughey 1960a, b, Iwan 1961, 1965, Roberts and Spanos 1990, Liu 2000). In the approach explained in this book, the following motivation was drawn based on the observation that the main part of a

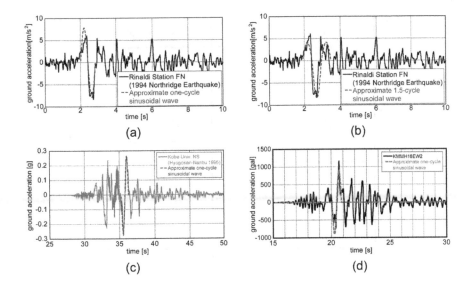

Figure 1.1 Recorded ground motions with pulse-type main parts modeled by simple sinusoidal wavelets, (a) Rinaldi Station FN (Northridge 1994) vs one-cycle sine wave, (b) Rinaldi Station FN (Northridge 1994) vs 1.5-cycle sine wave, (c) Kobe Univ. NS (Hyogoken-Nanbu 1995) vs one-cycle sine wave, (d) Mashiki EW (Kumamoto 2016, April 16) vs one-cycle sine wave.

near-fault ground motion is usually characterized by a one-cycle or a few-cycle sinusoidal wave as shown in Figure 1.1, and this part greatly influences the maximum deformation of building structures. If such one-cycle or a few-cycle sinusoidal wave as the main part of the near-fault ground motion can be represented by a double impulse as shown in Figure 1.2, the elastic-plastic response (continuation of free-vibrations) can be derived by using the full advantage of an energy balance approach without solving the equation of motion directly. The input of impulse is expressed by the instantaneous change of velocity of the structural mass leading to the instantaneous input of kinetic energy.

In the earthquake-resistant design, the resonance and the role of damping are two key issues, and they have been investigated extensively. In particular, since the resonance brings worse and critical effects to structures, it has been treated as a main theme in the earthquake-resistant design of structures. While the resonant equivalent frequency has to be computed for a specified input level by changing the excitation frequency in a parametric manner in case of treating the sinusoidal input (Caughey 1960a, b, Iwan 1961, 1965, Roberts and Spanos 1990, Liu 2000), no iteration is required in the method for the double impulse explained in this book. This is because the resonant equivalent frequency can be obtained directly without repetitive procedure.

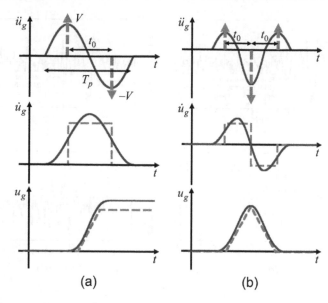

Figure 1.2 Simplification of ground motion (acceleration, velocity, displacement):
(a) Fling-step input and double impulse, (b) Forward-directivity input
and triple impulse (Kojima and Takewaki 2015).

In the double impulse, the analysis can be done without the input frequency
(timing of impulses) before the second impulse is inputted. The resonance
can be proved by using energy investigation, and the critical timing of the
second impulse can be characterized as the time with zero restoring force.
The maximum elastic-plastic response after impulse can be obtained by
equating the initial kinetic energy computed by the initial velocity to the
sum of hysteretic and elastic strain energies. It should be pointed out that
only critical response (upper bound) is captured by the method explained in
this book, and the critical resonant frequency can be obtained automatically
for the increasing input level of the double impulse.

In the history of the seismic-resistant design of building structures, the
earthquake input energy has played an important role together with deformation and acceleration (for example, Housner 1959, 1975, Berg and
Thomaides 1960, Housner and Jennings 1975, Zahrah and Hall 1984,
Akiyama 1985, Leger and Dussault 1992). While deformation and acceleration can predict and evaluate the local performance of a building structure
mainly for serviceability, the energy can evaluate the global performance of
a building structure mainly for safety. Especially energy is appropriate for
describing the performance of building structures of different sizes in a uni-
fied manner because energy is a global index different from deformation and
acceleration as local indices. In fact, in Japan, there are three criteria in
parallel: force, deformation, and energy. In 1981, the force was introduced

as a criterion for safety, and in 2000, the deformation was introduced as a criterion for safety. More recently, in 2005, the input energy evaluated from the design velocity response spectrum was used as a criterion. These three criteria are used now in parallel (BSL in Japan 1981, 2000, 2005).

A theory of earthquake input energy to building structures under single impulse was shown to be useful for disclosing the property of the energy transfer function (Takewaki 2004, 2007). This property means that the area of the energy transfer function is constant. The property of the energy transfer function similar to the case of a simple single-degree-of-freedom (SDOF) model has also been clarified for a swaying-rocking model. By using this property, the mechanism of earthquake input energy to the swaying-rocking model including the soil amplification has been made clear under the input of single impulse (Kojima et al. 2015b). However single impulse may be unrealistic because the frequency characteristic of input cannot be expressed by this input. In order to resolve such an issue, the double impulse is introduced in this book. Furthermore, because the elastic-plastic response is treated, the time-domain formulation is introduced in this book (Kojima and Takewaki 2015).

1.2 DOUBLE IMPULSE AND CORRESPONDING ONE-CYCLE SINE WAVE WITH THE SAME FREQUENCY AND SAME MAXIMUM FOURIER AMPLITUDE

The velocity amplitude V of the double impulse is related to the maximum velocity of the corresponding one-cycle sine wave with the same frequency (the period is twice the interval of the double impulse) so that the maximum Fourier amplitudes of both inputs coincide (Kojima 2018, Akehashi et al. 2018). The detail is explained in this section.

The double impulse is expressed by

$$\ddot{u}_g(t) = V\delta(t) - V\delta(t - t_0) \qquad (1.1)$$

where V is the velocity amplitude of the double impulse and $\delta(t)$ is the Dirac delta function. The Fourier transform of Eq. (1.1) can be obtained as

$$\ddot{U}_g(\omega) = V\left(1 - e^{-i\omega t_0}\right) \qquad (1.2)$$

Let A_p, T_p, $\omega_p = 2\pi/T_p$ denote the acceleration amplitude, the period and the circular frequency of the corresponding one-cycle sine wave, respectively. The acceleration wave $\ddot{u}_g^{SM}$ of the corresponding one-cycle sine wave is expressed by

$$\ddot{u}_g^{SM} = A_p \sin(\omega_p t) \quad \left(0 \le t \le T_p = 2t_0\right) \qquad (1.3)$$

The time interval t_0 of two impulses in the double impulse is related to the period T_p of the corresponding one-cycle sine wave by $T_p = 2t_0$. Although the starting points of both inputs differ by $t_0/2$, the starting time of one-cycle sine wave does not affect the Fourier amplitude. For this reason, the starting time of the one-cycle sine wave will be adjusted so that the responses of both inputs correspond well. In this section, the relation of the velocity amplitude V of the double impulse with the acceleration amplitude A_p of the corresponding one-cycle sine wave is derived. The ratio a of A_p to V is introduced by

$$A_p = aV \tag{1.4}$$

The Fourier transform of $\ddot{u}_g^{SM}$ in Eq. (1.3) is computed by

$$\ddot{U}_g^{SM}(\omega) = \int_0^{2t_0} \left\{ A_p \sin(\omega_p t) \right\} e^{-i\omega t} dt = \frac{\pi t_0 A_p}{\pi^2 - (\omega t_0)^2} \left(1 - e^{-2t_0 \omega i} \right) \tag{1.5}$$

From Eqs. (1.2) and (1.5), the Fourier amplitudes of both inputs are expressed by

$$\left| \ddot{U}_g(\omega) \right| = V \sqrt{2 - 2\cos(\omega t_0)} \tag{1.6}$$

$$\left| \ddot{U}_g^{SW}(\omega) \right| = A_p \left| \frac{2\pi t_0}{\pi^2 - (\omega t_0)^2} \sin(\omega t_0) \right| \tag{1.7}$$

The coefficient a can be derived from Eqs. (1.4), (1.6), (1.7), and the equivalence of the maximum Fourier amplitude $\left| \ddot{U}_g(\omega) \right|_{\max} = \left| \ddot{U}_g^{SW}(\omega) \right|_{\max}$.

$$a(t_0) = \frac{A_p}{V} = \frac{\max \left| \sqrt{2 - 2\cos(\omega t_0)} \right|}{\max \left| \dfrac{2\pi t_0}{\pi^2 - (\omega t_0)^2} \sin(\omega t_0) \right|} \tag{1.8}$$

In the numerator of Eq. (1.8), $\max \left| \sqrt{2 - 2\cos(\omega t_0)} \right| = 2$ holds. The denominator $\max |2\pi t_0 \sin(\omega t_0)/\{\pi^2 - (\omega t_0)^2\}|$ in Eq. (1.8) will be evaluated next. Let us define the function $f(x)$ given by

$$f(x) = \frac{1}{\pi^2 - x^2} \sin x \tag{1.9}$$

The maximum value $f_{\max}$ of $f(x)$ and the corresponding argument $x = x_0$ can be obtained as follows.

$$x_0 = 2.63099585... \qquad (1.10)$$

$$f_{max} = f(x = x_0) = 0.165802809... \qquad (1.11)$$

The values in Eqs. (1.10) and (1.11) were obtained numerically. From Eqs. (1.8)–(1.11), the coefficient a is expressed as a function of the time interval t_0 of two impulses.

$$a(t_0) = 1/(\pi t_0 f_{max}) \qquad (1.12)$$

Figure 1.3(a) shows the relation between t_0 and a. Furthermore, Figure 1.3(b) presents examples of the Fourier amplitudes of both inputs with the same maximum Fourier amplitude. Since the Fourier amplitudes of both inputs differ greatly in larger frequencies, further investigation will be necessary in dealing with multi-degree-of-freedom models.

Consider next the ratio of the maximum velocity V_p of the one-cycle sine wave to the velocity amplitude V of the double impulse. The velocity $\dot{u}_g^{SW}$ of the one-cycle acceleration sine wave is expressed by

$$\dot{u}_g^{SW} = \int_0^t \ddot{u}_g^{SW} dt = \int_0^t A_p \sin(\omega_p t) dt = \frac{A_p}{\omega_p}\{1 - \cos(\omega_p t)\} \qquad (1.13)$$

From Eq. (1.13), the maximum velocity V_p of the one-cycle sine wave can be expressed by

$$V_p = 2A_p/\omega_p \qquad (1.14)$$

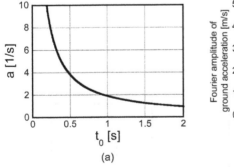
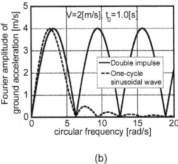

(a)

(b)

Figure 1.3 Relation of amplitudes between double impulse and one-cycle sine wave, (a) Coefficient a (=A_P/V) with respect to impulse timing t_0, (b) Fourier amplitude of double impulse and one-cycle sine wave (Akehashi et al. 2018).

Eqs. (1.4), (1.12), (1.14), and $\omega_p = \pi/t_0$ lead to the relation between V_p and V.

$$V_p = \left\{ 2/\left(\pi^2 f_{\max}\right) \right\} V \qquad (1.15)$$

From Eqs. (1.11) and (1.15), V_p/V is expressed as

$$V_p/V = 2/\left(\pi^2 f_{\max}\right) = 1.22218898... \qquad (1.16)$$

It can be found from Eq. (1.16) that if the maximum Fourier amplitudes of both inputs are the same, the ratio of V_p to V becomes constant. The modulated one-cycle sine wave will be called "the corresponding one-cycle sine wave."

1.3 ENERGY BALANCE UNDER EARTHQUAKE GROUND MOTION AND IMPULSE

The energy balance under an earthquake ground motion and the corresponding impulse, which plays a central role in this book, is explained in this section.

1.3.1 Undamped model

Consider a single-degree-of-freedom (SDOF) model of mass m and stiffness k subjected to an earthquake ground acceleration $\ddot{u}_g(t)$ (see Figure 1.4(a)). Assume the initial condition $x(0) = \dot{x}(0) = 0$. Let $x(t)$ denote the displacement of the mass relative to the ground. The equation of motion of the model can be described by

$$m\ddot{x} + kx = -m\ddot{u}_g \qquad (1.17)$$

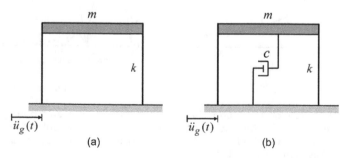

Figure 1.4 SDOF model, (a) Undamped model, (b) Damped model.

Let's consider the vibration of the model during $t = [0, t_0]$. Multiplication of the velocity $\dot{x}$ on both sides of Eq. (1.17) and integration over $t = [0, t_0]$ provide

$$\left[(1/2)m\dot{x}(t)^2 \right]_0^{t_0} + \left[(1/2)kx(t)^2 \right]_0^{t_0} = -\int_0^{t_0} m\ddot{u}_g(t)\dot{x}(t)\,dt \qquad (1.18)$$

With the initial condition, Eq. (1.18) leads to

$$(1/2)m\dot{x}(t_0)^2 + (1/2)kx(t_0)^2 = -\int_0^{t_0} m\ddot{u}_g(t)\dot{x}(t)\,dt \qquad (1.19)$$

The first term of the left-hand side of Eq. (1.19) indicates the kinetic energy and the second term is the strain energy. It is important to note that, only after the response $x(t)$ is computed numerically, the energy balance can be evaluated.

On the other hand, consider that the model is subjected to the single impulse $\ddot{u}_g(t) = V\delta(t)$. In this case, since $x(0) = 0, \dot{x}(0) = -V$, Eq. (1.18) can be expressed as

$$(1/2)m\dot{x}(t_0)^2 - (1/2)mV^2 + (1/2)kx(t_0)^2 = 0 \qquad (1.20)$$

Assume that the maximum displacement occurs at $t = t_0$ (see Figure 1.5(a)). Since $\dot{x}(t_0) = 0$, Eq. (1.20) yields

$$(1/2)kx(t_0)^2 = (1/2)mV^2 \qquad (1.21)$$

Eq. (1.21) means that the maximum displacement can be obtained from the energy balance law.

This principle can be applied to an elastic-plastic model. Let $f(x)$ denote the nonlinear restoring-force characteristic of the elastic-plastic model. For this model, Eq. (1.20) can be modified into

$$(1/2)m\dot{x}(t_0)^2 - (1/2)mV^2 + \int_0^{x_0} f(x)\,dx = 0 \qquad (1.22)$$

where $x_0 = x(t_0)$. Assuming again that the maximum displacement occurs at $t = t_0$, Eq. (1.22) leads to

$$\int_0^{x_0} f(x)\,dx = (1/2)mV^2 \qquad (1.23)$$

By specifying an explicit expression of $f(x)$, e.g., an elastic-perfectly plastic one (see Figure 1.5(b)) or a bilinear one, Eq. (1.23) provides the maximum displacement x_0.

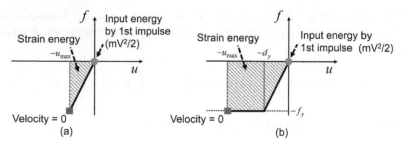

Figure 1.5 Input energy (kinetic energy) by impulse and strain energy, (a) Elastic model, (b) Elastic-plastic model.

1.3.2 Damped model

Consider next a SDOF model of mass m, stiffness k, viscous damping coefficient c subjected to an earthquake ground acceleration $\ddot{u}_g(t)$ (see Figure 1.4(b)). Assume again the initial condition $x(0) = \dot{x}(0) = 0$. The equation of motion of this model can be expressed by

$$m\ddot{x} + c\dot{x} + kx = -m\ddot{u}_g \qquad (1.24)$$

Consider the vibration of the model during $t = [0, t_0]$. Multiplication of the velocity $\dot{x}$ on both sides of Eq. (1.24) and integration over $t = [0, t_0]$ provide

$$\left[(1/2)m\dot{x}(t)^2\right]_0^{t_0} + \int_0^{t_0} c\dot{x}(t)^2\, dt + \left[(1/2)kx(t)^2\right]_0^{t_0} = -\int_0^{t_0} m\ddot{u}_g(t)\dot{x}(t)\, dt \qquad (1.25)$$

Application of the initial condition to Eq. (1.25) leads to

$$(1/2)m\dot{x}(t_0)^2 + \int_0^{t_0} c\dot{x}(t)^2\, dt + (1/2)kx(t_0)^2 = -\int_0^{t_0} m\ddot{u}_g(t)\dot{x}(t)\, dt \qquad (1.26)$$

The first term of the left-hand side of Eq. (1.26) indicates the kinetic energy and the third term is the strain energy. The second term is the energy dissipated by the viscous damping. As in the undamped case, it should be remarked that only after the response $x(t)$ is computed numerically can the energy balance be evaluated.

On the other hand, consider that the model is subjected to the single impulse $\ddot{u}_g(t) = V\delta(t)$. In this case, since $x(0) = 0, \dot{x}(0) = -V$, Eq. (1.26) can be expressed as

$$(1/2)m\dot{x}(t_0)^2 - (1/2)mV^2 + \int_0^{t_0} c\dot{x}(t)^2\, dt + (1/2)kx(t_0)^2 = 0 \qquad (1.27)$$

Assume that the maximum displacement occurs at $t = t_0$. Since $\dot{x}(t_0) = 0$, Eq. (1.27) yields

$$(1/2)kx(t_0)^2 = (1/2)mV^2 - \int_0^{t_0} c\dot{x}(t)^2 \, dt \qquad (1.28)$$

Eq. (1.28) indicates that the maximum displacement can be obtained from the energy balance law. However, the energy dissipated by the viscous damping has to be evaluated in an appropriate manner in this case. This principle can also be applied to an elastic-plastic model. The left-hand side of Eq. (1.28) can be replaced by $\int_0^{x_0} f(x) \, dx$. If the second term of the right-hand side of Eq. (1.28) can be expressed approximately in terms of x_0, x_0 can be evaluated by Eq. (1.28) after replacing the left-hand side of Eq. (1.28) by the term $\int_0^{x_0} f(x) \, dx$.

It can be shown that the right-hand side of Eq. (1.25) expresses the total input energy to the model. The work done by the ground input on the SDOF model (see Figure 1.6) can be expressed by

$$\int_0^{x_0} m\{\ddot{u}_g(t) + \ddot{x}(t)\} \, dx = \int_0^{t_0} m\{\ddot{u}_g(t) + \ddot{x}(t)\}\dot{x}(t) \, dt \qquad (1.29)$$

Integration by parts of Eq. (1.29) leads to

$$\int_0^{t_0} m\{\ddot{u}_g(t) + \ddot{x}(t)\}\dot{x}(t) \, dt = \left[(1/2)m\dot{x}\dot{u}_g\right]_0^{t_0} - \int_0^{t_0} m\dot{x}\ddot{u}_g \, dt + \left[(1/2)m\dot{u}_g^2\right]_0^{t_0} \qquad (1.30)$$

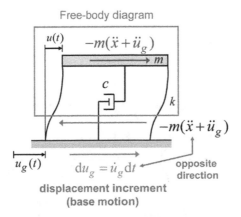

Figure 1.6 SDOF model subjected to earthquake ground motion (Kojima et al. 2015a).

The condition $\dot{u}_g(0) = \dot{u}_g(t_0) = 0$ provides

$$\int_0^{t_0} m\{\ddot{u}_g(t) + \ddot{x}(t)\}\dot{x}(t)\,dt = -\int_0^{t_0} m\dot{x}\ddot{u}_g\,dt \qquad (1.31)$$

This means that, if t_0 indicates the final time of input ground motion, the right-hand side of Eq. (1.25) expresses the total input energy to the model.

1.4 CRITICAL INPUT TIMING OF SECOND IMPULSE IN DOUBLE IMPULSE

The energy balance approach explained in the above section is often used in the seismic-resistant design of structures. A more important aspect in the present approach is the critical timing of the second impulse in the double impulse input. The energy balance law explained in the above section (after the input of the first impulse until the maximum displacement) can also be used in a similar manner for the response process after the second impulse until the next (inverse-direction) maximum displacement. The initial velocity V just after the first impulse has to be changed to $v_{max}^* + V$ where v_{max}^* is the velocity of mass just before the second impulse is applied (see Figure 1.7). Because the velocity attains the maximum at the point of zero restoring force, the subscript "max" is given. Since it can be proved that the critical timing of the second impulse is the time when the restoring force attains zero (see Chapter 2), the area of the nonlinear restoring-force characteristic from the zero restoring force (applied point of the second impulse) to the next maximum displacement u_{max2} can be evaluated without difficulty. A more detailed explanation will be provided in Chapter 2.

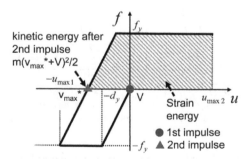

Figure 1.7 Critical input timing of second impulse attaining maximum value of u_{max2}.

Table 1.1 Comparison of conventional methods and proposed method for nonlinear resonant analysis.

Conventional method 1960, 1961 Caughey, Iwan	Proposed method 2015 Kojima and Takewaki
① Steady state	① Transient and steady state
② Impossible for elastic–perfectly plastic	② Possible even for elastic–perfectly plastic (any bilinear)
③ Repetition required (equivalent parameters/ resonant frequency)	③ No repetition required
	<Proposed method enables>
	→Closed-form **critical** response of elastic-plastic structure
afterward, stochastic linearization(transient response)	→Derive **resonant frequency** (impulse interval) without repetition
	→Closed-form **noncritical** response of elastic-plastic structure based on closed-form **critical** response

1.5 COMPARISON OF CONVENTIONAL METHODS AND THE PROPOSED METHOD FOR NONLINEAR RESONANT ANALYSIS

Table 1.1 shows the comparison of conventional methods and the proposed method for nonlinear resonant analysis. In the conventional methods represented by Caughey (1960a, b) and Iwan (1965), a steady state was treated and the analysis of the elastic–perfectly plastic model was impossible because of the computational stability. In addition, repetition was required for obtaining the convergent values or solving the transcendental equations. On the other hand, in the proposed method, both transient and steady-state responses can be dealt with. In addition, the analysis of the elastic–perfectly plastic model is possible and no iteration is required. Furthermore, the proposed method enables (1) the derivation of closed-form critical response of elastic-plastic structures, (2) the capture of the resonant frequency without repetition, and (3) the derivation of closed-form expressions on the noncritical responses of elastic-plastic structures. The detail of these facts will be explained in subsequent chapters.

1.6 OUTLINE OF THIS BOOK

In Chapter 1, the motivation of the proposed earthquake energy balance approach using impulses is explained. The simplification of fling-step near-fault ground motions into the double impulse is explained and the earthquake energy balance approach is introduced for undamped elastic–perfectly plastic (EPP) single-degree-of-freedom (SDOF) models under the critical double impulse.

In Chapter 2, a closed-form expression is derived by using the energy balance law for the maximum response of an undamped EPP SDOF model under the critical double impulse input.

In Chapter 3, a closed-form expression is derived by using the energy balance law for the critical response of an undamped EPP SDOF model under the triple impulse input as a representative of forward-directivity near-fault ground motions. Complicated phenomena on the critical response under the triple impulse input are investigated. The existence of the third impulse brings such complicated phenomena on the critical response.

In Chapter 4, the multiple impulse input of equal time interval is introduced as a substitute of long-duration earthquake ground motions which is expressed in terms of harmonic waves. A closed-form expression is derived by using the energy balance law for the maximum response of an EPP SDOF model under the critical multiple impulse input.

In Chapter 5, the double impulse is introduced as a good substitute for the one-cycle sinusoidal wave in representing the main part of a near-fault ground motion. A closed-form expression is derived by using the energy balance law for the maximum deformation of an EPP SDOF model with viscous damping under the critical double impulse. It uses (1) a quadratic function to approximate the damping force-deformation relation, (2) the assumption that the zero restoring-force timing in the unloading process is the critical timing of the second impulse, and (3) the energy balance law for the elastic-plastic system with viscous damping.

In Chapter 6, the multi-impulse is introduced as a substitute of the long-duration ground motion and the closed-form expression is derived for the steady-state elastic-plastic response of a bilinear hysteretic SDOF system under the critical multi-impulse. While the computation of the resonant equivalent frequency of the elastic-plastic system is a tough task in the conventional method dealing directly with the sinusoidal wave (Iwan 1961), it is shown that the steady-state elastic-plastic response under the critical multi-impulse can be obtained in closed form (without repetition) by using the energy balance law and the critical time interval of the multi-impulse (the resonant frequency) can also be obtained in closed form for the increasing input level.

In Chapter 7, the double impulse is introduced as a substitute of the fling-step near-fault ground motion. A closed-form expression for the elastic-plastic response of an EPP SDOF model on the compliant (flexible) ground by the "critical double impulse" is derived based on the expression for the corresponding EPP SDOF model with a fixed base. It is shown that the closed-form expression for the critical elastic-plastic response of the superstructure enables the clarification of the relation of the critical elastic-plastic response of the superstructure with the ground stiffness.

In Chapter 8, the closed-form expression for the maximum elastic-plastic response of a bilinear hysteretic SDOF model under the critical double impulse (Kojima and Takewaki 2016a) is extended to a dynamic stability (collapse) problem of elastic-plastic SDOF models with negative post-yield stiffness. Negative post-yield stiffness is treated to consider the P-delta effect. The double impulse is used as a substitute for the fling-step near-fault ground motion. The dynamic stability (collapse) limit of the velocity level is obtained for the critical input case by using the energy balance law.

In Chapter 9, an explicit limit on the input velocity level of the double impulse as a representative of the principal part of a near-fault ground motion is derived for the overturning of a rigid block. The energy balance law and the conservation law of angular momenta of the rigid block are used for describing and determining the rocking response under the critical double impulse.

In Chapter 10, because the response of 2DOF elastic-plastic building structures is quite complicated due to the phase lag between two masses compared to SDOF models for which a closed-form critical response can be derived, the upper bound of the critical response is introduced by using the convex model. The accuracy of the upper bound is then investigated.

In Chapter 11, an innovative method for optimal viscous damper placement is explained for EPP multi-degree-of-freedom (MDOF) shear building structures subjected to the critical double impulse as a representative of near-fault ground motions. Simultaneous treatment of nonlinear MDOF structures and uncertainty in the selection of input is the most remarkable point that has never been overcome in the past.

In Chapter 12, some future directions are explained which include

1. Treatment of noncritical case
2. Extension to nonlinear viscous damper and hysteretic damper
3. Treatment of uncertain fault-rupture model and uncertain deep ground property
4. Application to passive control systems for practical tall buildings
5. Stopper system for pulse-type ground motion of extremely large amplitude
6. Repeated single impulse in the same direction for repetitive ground motion input
7. Robustness evaluation
8. Principles in seismic resistant design (constant energy law, constant displacement law, law for resonant case)

In particular, application of the proposed approach to more practical situations will certainly enhance the broad and profound significance of the proposed innovative approach to earthquake structural engineering in nonlinear structural dynamics.

1.7 SUMMARIES

The obtained results may be summarized as follows:

1. It is possible to capture the intrinsic characters of near-fault pulse-type ground motions by replacing those by the double impulse input with the equivalent maximum Fourier amplitude. It can be a good substitute of the fling-step near-fault ground motion and a closed-form solution can be derived in the elastic-plastic response of a structure under the critical double impulse input by using the energy balance law.
2. Since only the free-vibration is induced in such double impulse input, the energy balance approach plays an important role in the derivation of the closed-form solution of a complicated elastic-plastic response. In other words, the energy balance approach enables the derivation of the maximum elastic-plastic seismic response without resorting to solving the differential equation. In this process, the input of impulse is expressed by the instantaneous change of velocity of the structural mass. The maximum elastic-plastic response after impulse can be obtained through the energy balance law by equating the initial kinetic energy computed by the initial velocity to the sum of hysteretic and elastic strain energies.

REFERENCES

Abbas, A. M., Manohar, C. S. (2002). Investigations into critical earthquake load models within deterministic and probabilistic frameworks, *Earthquake Eng. Struct. Dyn.*, 31, 813–832.

Abrahamson, N., Ashford, S., Elgamal, A., Kramer, S., Seible, F., Sommerville, P. (1998). *Proc. of First PEER Workshop on Characterization of Specific Source Effects.*

Akehashi, H., Kojima, K. and Takewaki, I. (2018). Critical response of SDOF damped bilinear hysteretic system under double impulse as substitute for near-fault ground motion, *Frontiers in Built Environment*, 4: 5.

Akiyama, H. (1985). *Earthquake resistant limit-state design for buildings.* University of Tokyo Press, Tokyo, Japan.

Alavi, B. and Krawinkler, H. (2004). Behaviour of moment resisting frame structures subjected to near-fault ground motions, *Earthquake Eng. Struct. Dyn.*, 33(6), 687–706.

Berg, G. V., and Thomaides T. T. (1960). "Energy consumption by structures in strong-motion earthquakes," in *Proceedings of 2nd World Conference on Earthquake Engineering*, Tokyo and Kyoto, 681–696.

Bertero, V. V., Mahin, S. A., and Herrera, R. A. (1978). Aseismic design implications of near-fault San Fernando earthquake records, *Earthquake Eng. Struct. Dyn.*, 6(1), 31–42.

Bozorgnia, Y. and Campbell, K. W. (2004). Engineering characterization of ground motion, Chapter 5 in *Earthquake engineering from engineering seismology to performance-based engineering*, (eds.) Bozorgnia, Y. and Bertero, V. V., CRC Press, Boca Raton.

Bray, J. D. and Rodriguez-Marek, A. (2004). Characterization of forward-directivity ground motions in the near-fault region, *Soil Dyn. Earthquake Eng.*, 24(11), 815–828.

Building Standard Law in Japan: Earthquake resistant design code (1981, 2000, 2005).

Caughey, T.K. (1960a). Sinusoidal excitation of a system with bilinear hysteresis, *J. Appl. Mech.*, 27(4), 640–643.

Caughey, T. K. (1960b). Random excitation of a system with bilinear hysteresis, *J. Appl. Mech.*, 27(4), 649–652.

Drenick, R. F. (1970). Model-free design of aseismic structures. *J. Eng. Mech. Div.*, ASCE, 96(EM4), 483–493.

Fukumoto, Y. and Takewaki, I. (2017). Dual control high-rise building for robuster earthquake performance, *Frontiers in Built Environment* (Specialty Section: Earthquake Engineering), Volume 3: Article 12.

Hall, J. F., Heaton, T. H., Halling, M. W., and Wald, D. J. (1995). Near-source ground motion and its effects on flexible buildings, *Earthuake Spectra*, 11(4), 569–605.

Hayden, C. P., Bray, J. D., and Abrahamson, N. A. (2014). Selection of near-fault pulse motions, *J. Geotechnical Geoenvironmental Eng.*, ASCE, 140(7).

Housner, G. W. (1959). Behavior of structures during earthquakes, *J. Eng. Mech. Div.*, ASCE, 85(4), 109–129.

Housner, G. W. (1975). "Measures of severity of earthquake ground shaking," in *Proceedings of the US National Conference on Earthquake Engineering*, Ann Arbor, Michigan, 25–33.

Housner, G. W., and Jennings, P. C. (1975). "The capacity of extreme earthquake motions to damage structures," in *Structural and geotechnical mechanics*: A volume honoring N.M. Newmark edited by W.J. Hall, 102–116, Prentice-Hall Englewood Cliff, NJ.

Iwan, W. D. (1961). *The dynamic response of bilinear hysteretic systems*, Ph.D. Thesis, California Institute of Technology.

Iwan, W. D. (1965). The dynamic response of the one-degree-of-freedom bilinear hysteretic system, *Proceedings of the Third World Conference on Earthquake Engineering*, 1965.

Kalkan, E., and Kunnath, S. K. (2006). Effects of fling step and forward directivity on seismic response of buildings, *Earthquake Spectra*, 22(2), 367–390.

Kalkan, E., and Kunnath, S. K. (2007). Effective cyclic energy as a measure of seismic demand, *J. Earthquake Eng.*, 11, 725–751.

Kanno, Y., and Takewaki, I. (2016). Robustness analysis of elastoplastic structure subjected to double impulse, *J. of Sound and Vibration*, 383, 309–323.

Kanno, Y., Yasuda, K., Fujita, K., and Takewaki, I. (2017). Robustness of SDOF elastoplastic structure subjected to double-impulse input under simultaneous uncertainties of yield deformation and stiffness, *Int. J. Non-Linear Mechanics*, 91, 151–162.

Khaloo, A. R., Khosravi, H., and Hamidi Jamnani, H. (2015). Nonlinear interstory drift contours for idealized forward directivity pulses using "Modified Fish-Bone" models, *Advances in Structural Eng.*,18(5), 603–627.

Kojima, K. (2018). Earthquake critical excitation method for elastic-plastic structures using impulse sequence and energy balance law, Ph.D. Thesis, Graduate School of Engineering, Kyoto University (in Japanese).

Kojima, K., Fujita, K., and Takewaki, I. (2015a). Critical double impulse input and bound of earthquake input energy to building structure, *Frontiers in Built Environment*, Volume 1, Article 5.

Kojima, K., Sakaguchi, K., and Takewaki, I. (2015b). Mechanism and bounding of earthquake energy input to building structure on surface ground subjected to engineering bedrock motion, *Soil Dyn. Earthquake Eng.*, 70, 93–103.

Kojima, K., Saotome, Y., and Takewaki, I. (2017). Critical earthquake response of SDOF elastic-perfectly plastic model with viscous damping under double impulse as substitute of near-fault ground motion, *J. of Structural and Construction Eng.* (AIJ), 82(735), 643–652 (in Japanese).

Kojima, K., and Takewaki, I. (2015). Critical earthquake response of elastic-plastic structures under near-fault ground motions (Part 1: Fling-step input), *Frontiers in Built Environment* (Specialty Section: Earthquake Engineering), Volume 1, Article 12.

Kojima, K., and Takewaki, I. (2016a). Closed-form critical earthquake response of elastic-plastic structures with bilinear hysteresis under near-fault ground motions, *J. of Structural and Construction Eng.* (AIJ), 81(726), 1209–1219 (in Japanese).

Kojima, K., and Takewaki, I. (2016b). Closed-form critical earthquake response of elastic-plastic structures on compliant ground under near-fault ground motions, *Frontiers in Built Environment* (Specialty Section: Earthquake Engineering), Volume 2, Article 1.

Kojima, K., and Takewaki, I. (2016c). Closed-form dynamic stability criterion for elastic-plastic structures under near-fault ground motions, *Frontiers in Built Environment* (Specialty Section: Earthquake Engineering), Volume 2, Article 6.

Kojima, K., and Takewaki, I. (2016d). A simple evaluation method of seismic resistance of residential house under two consecutive severe ground motions with intensity 7, *Frontiers in Built Environment* (Specialty Section: Earthquake Engineering) , Volume 2, Article 15.

Leger, P., and Dussault, S. (1992). Seismic-energy dissipation in MDOF structures, *J. Struct. Eng.*, ASCE, 118(5), 1251–1269.

Liu C.-S. (2000). The steady loops of SDOF perfectly elastoplastic structures under sinusoidal loadings, *J. Marine Science and Technology*, 8(1), 50–60.

Mavroeidis, G. P., Dong, G., and Papageorgiou, A. S. (2004). Near-fault ground motions, and the response of elastic and inelastic single-degree-freedom (SDOF) systems, *Earthquake Eng. Struct. Dyn.*, 33, 1023–1049.

Mavroeidis, G. P., and Papageorgiou, A. S. (2003). A mathematical representation of near-fault ground motions, *Bull. Seism. Soc. Am.*, 93(3), 1099–1131.

Moustafa, A., Ueno, K., and Takewaki, I. (2010). Critical earthquake loads for SDOF inelastic structures considering evolution of seismic waves, *Earthquakes and Structures*, 1(2), 147–162.

Mukhopadhyay, S., and Gupta, V. K. (2013a). Directivity pulses in near-fault ground motions—I: Identification, extraction and modeling, *Soil Dyn. Earthquake Eng.*, 50, 1–15.

Mukhopadhyay, S., and Gupta, V. K. (2013b). Directivity pulses in near-fault ground motions—II: Estimation of pulse parameters, *Soil Dyn. Earthquake Eng.*, 50, 38–52.

Nabeshima, K., Taniguchi, R., Kojima, K., and Takewaki, I. (2016). Closed-form overturning limit of rigid block under critical near-fault ground motions, *Frontiers in Built Environment* (Specialty Section: Earthquake Engineering), Volume 2, Article 9.

PEER Center (2013). *PEER ground motion database*, Pacific Earthquake Engineering Research Center, Richmond, CA.

PEER Center, ATC, Japan Ministry of education, science, sports, and culture, US-NSF (2000). Effects of near-field earthquake shaking, *Proc. of US-Japan workshop*, San Francisco, March 20–21, 2000.

Roberts, J. B., and Spanos, P. D. (1990). *Random vibration and statistical linearization*. Wiley, New York.

Rupakhety, R., and Sigbjörnsson, R. (2011). Can simple pulses adequately represent near-fault ground motions?, *J. Earthquake Eng.*, 15,1260–1272.

Sasani, M., and Bertero, V. V. (2000). "Importance of severe pulse-type ground motions in performance-based engineering: historical and critical review," in *Proceedings of the Twelfth World Conference on Earthquake Engineering*, Auckland, New Zealand.

Shiomi, T., Fujita, K., Tsuji, M., and Takewaki, I. (2016). Explicit optimal hysteretic damper design in elastic-plastic structure under double impulse as representative of near-fault ground motion, *Int. J. Earthquake and Impact Engineering*, 1(1/2), 5–19.

Takewaki, I. (1998). Resonance and criticality measure of ground motions via probabilistic critical excitation method, *Soil Dyn. Earthquake Eng.*, 21, 645–659.

Takewaki, I. (2002). Robust building stiffness design for variable critical excitations, *J. Struct. Eng.*, ASCE, 128(12), 1565–1574.

Takewaki, I. (2004). Bound of earthquake input energy, *J. Struct. Eng.*, ASCE, 130(9), 1289–1297.

Takewaki, I. (2007). *Critical excitation methods in earthquake engineering*, Elsevier, Second edition in 2013.

Takewaki, I., Moustafa, A. and Fujita, K. (2012). *Improving the earthquake resilience of buildings: The worst case approach*, Springer, London.

Takewaki, I., and Tsujimoto, H. (2011). Scaling of design earthquake ground motions for tall buildings based on drift and input energy demands, *Earthquakes Struct.*, 2(2), 171–187.

Taniguchi, R., Kojima, K., and Takewaki, I. (2016). Critical response of 2DOF elastic-plastic building structures under double impulse as substitute of near-fault ground motion, *Frontiers in Built Environment* (Specialty Section: Earthquake Engineering), Volume 2, Article 2.

Taniguchi, R., Nabeshima, K., Kojima, K., and Takewaki, I. (2017). Closed-form rocking vibration of rigid block under critical and non-critical double impulse, *Int. J. Earthquake and Impact Engineering*, 2(1), 32–45.

Vafaei, D., and Eskandari, R. (2015). Seismic response of mega buckling-restrained braces subjected to fling-step and forward-directivity near-fault ground motions, *Struct. Design Tall Spec. Build.*, 24, 672–686.

Xu, Z., Agrawal, A. K., He, W.-L., and Tan, P. (2007). Performance of passive energy dissipation systems during near-field ground motion type pulses, *Eng. Struct.*, 29, 224–236.

Yamamoto, K., Fujita, K., and Takewaki, I. (2011). Instantaneous earthquake input energy and sensitivity in base-isolated building, *Struct. Design Tall Spec. Build.*, 20(6), 631–648.

Yang, D., and Zhou, J. (2014). A stochastic model and synthesis for near-fault impulsive ground motions, *Earthquake Eng. Struct. Dyn.*, 44, 243–264.

Zahrah, T. F., and Hall, W. J. (1984). Earthquake energy absorption in SDOF structures. *J. Struct. Eng.*, ASCE, 110(8), 1757–1772.

Zhai, C., Chang, Z., Li, S., Chen, Z.-Q., and Xie, L. (2013). Quantitative identification of near-fault pulse-like ground motions based on energy, *Bull. Seism. Soc. Am.*, 103(5), 2591–2603.

Chapter 2

Critical earthquake response of an elastic–perfectly plastic SDOF model under double impulse as a representative of near-fault ground motions

2.1 INTRODUCTION

Critical excitation problems were posed independently by Dr. Drenick and Dr. Shinozuka in 1970 (Drenick 1970, Shinozuka 1970) and acknowledged as one of the important fields in applied mechanics. More detailed description on the essential features of the critical excitation problems can be found in Takewaki (2007), which is the first and unique monograph on critical excitation and includes its historical sketch. In contrast to the linear elastic response in the early stage, some critical excitation methods for elastic-plastic responses were treated comprehensively using the equivalent linearization methods in the reference (Takewaki 2007). Furthermore, deterministic critical excitation methods for elastic-plastic responses were dealt with by using the mathematical programming (SQP) in the same reference. In both approaches, repetitive computations were needed. Compared to these approaches, a more straightforward critical excitation method enabling the derivation of closed-form expressions on the critical elastic-plastic responses is explained in this chapter. This approach is based on an innovative concept in nonlinear structural dynamics, i.e., the substitution of earthquake ground motions by the impulse input and the capture of the critical timing of the impulse. This style is consistent throughout this book.

As pointed out in Chapter 1, there are two types of earthquake ground motions in general. One is the near-fault ground motion and the other is the long-period, long-duration ground motion. In this chapter, the former one is discussed, and the double impulse input is regarded as its simplification.

2.2 DOUBLE IMPULSE INPUT

Consider a ground acceleration $\ddot{u}_g(t)$ in terms of double impulse expressed by

$$\ddot{u}_g(t) = V\delta(t) - V\delta(t - t_0) \tag{2.1}$$

where V is the given initial velocity (velocity amplitude), $\delta(t)$ is the Dirac delta function, and t_0 is the time interval between two impulses.

2.3 SDOF SYSTEM

Consider an undamped elastic–perfectly plastic (EPP) single-degree-of-freedom (SDOF) system of mass m and stiffness k. The yield deformation and yield force are denoted by d_y and f_y, respectively (see Figure 2.1). Let $\omega_1 = \sqrt{k/m}$, u, and f denote the undamped natural circular frequency, the displacement of the mass relative to the ground, and the restoring force of the system, respectively. The time derivative is denoted by an over-dot. In Section 2.4, these parameters will be dealt with in a dimensionless or normalized form to derive the relation of permanent interest between the input and the elastic-plastic response. However, numerical parameters will be used partially in Section 2.5 to demonstrate an example setting of actual parameters.

2.4 MAXIMUM ELASTIC-PLASTIC DEFORMATION OF SDOF SYSTEM TO DOUBLE IMPULSE

The elastic-plastic response of the EPP SDOF system to the double impulse can be described by the continuation of free-vibrations with sudden velocity change in the mass at the action of impulse. Different from the linear elastic system, the superposition principle for both impulses cannot be used for the elastic-plastic system. The maximum deformation after the first impulse is denoted by u_{max1} and that after the second impulse is expressed by u_{max2} (u_{max1} and u_{max2} are the absolute value) as shown in Figure 2.1. It is noted that, if u_{max1} is large, u_{max2} may become smaller for the second impulse. This may be related to the drift of the origin in the restoring-force characteristic. The input of each impulse results in the instantaneous change of velocity of the structural mass. Such responses can be derived by an energy balance approach without resorting to solving the equation of motion directly. The kinetic energy given at the initial stage (at the time of action of the first impulse) and at the time of action of the second impulse is transformed into the sum of the hysteretic dissipation energy in terms of the plastic deformation and the strain energy in terms of the yield deformation. The corresponding schematic diagram was shown in Section 1.3 in Chapter 1. By using this rule, the maximum deformation can be obtained in a simple manner.

It should be emphasized that, while the resonant equivalent frequency can be computed for a specified input level by changing the excitation frequency parametrically in the conventional methods dealing with the sinusoidal input (Caughey 1960a, b, Roberts and Spanos 1990, Liu 2000, Moustafa

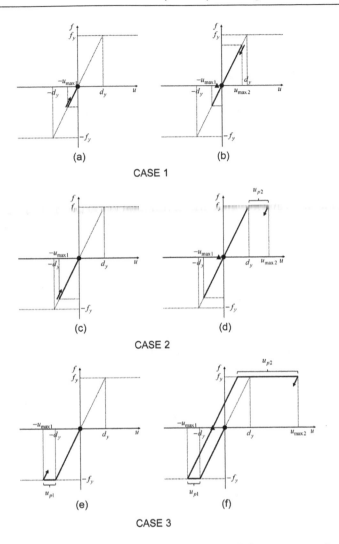

Figure 2.1 Prediction of maximum elastic-plastic deformation under double impulse based on energy approach: (a), (b) CASE 1: Elastic response; (c), (d) CASE 2: Plastic response after the second impulse; (e), (f) CASE 3: Plastic response after the first impulse (● : first impulse,▲ : second impulse) (Kojima and Takewaki 2015).

et al. 2010), no iteration is required in the method for the double impulse explained in this chapter. This is because the resonant equivalent frequency (resonance can be proved by using energy investigation: see Appendix 1) can be obtained directly without the repetitive procedure. As a result, the critical timing of the second impulse can be characterized as the time corresponding to zero restoring force.

While resonance curves were drawn in the conventional methods, only critical response (upper bound) is captured by the method explained in this chapter, and the critical resonant frequency can be obtained automatically for the increasing input level of the double impulse. One of the original points of the approach explained in this chapter is the introduction of the concept of "critical excitation" in the elastic-plastic response (Drenick 1970, Abbas and Manohar 2002, Takewaki 2007, Moustafa et al. 2010). Once the frequency and amplitude of the critical double impulse are computed, the corresponding one-cycle sinusoidal motion as a representative of the fling-step motion can be identified (see Section 1.2 in Chapter 1).

Let us explain the evaluation method of u_{max1} and u_{max2} in the following. The plastic deformation after the first impulse is expressed by u_{p1} and that after the second impulse is denoted by u_{p2} as shown in Figure 2.1. There are three cases to be considered depending on the yielding stage as shown in Figure 2.1. Let us introduce a new parameter $V_y(=\omega_1 d_y)$ for expressing the input level of the velocity of the double impulse at which the EPP SDOF system just attains the yield deformation after the first impulse. This parameter enables the nondimensional expression of the input velocity.

Let us refer to Figure 1.5(a) for the energy balance law. Figures 2.1(a) and 2.1(b) show the maximum deformations after the first and critical second impulses, respectively, for the elastic case (CASE 1) during the whole stage. The maximum deformation u_{max1} after the first impulse can be obtained from the following energy balance law (also conservation law in this case).

$$mV^2/2 = ku_{max1}^2/2 \qquad (2.2)$$

On the other hand, the maximum deformation u_{max2} after the critical second impulse can be computed from another energy balance law (also conservation law in this case).

$$m(2V)^2/2 = ku_{max2}^2/2 \qquad (2.3)$$

As explained earlier in Chapter 1, the critical timing of the second impulse is the time when the zero restoring force is attained in the unloading process. The velocity V induced by the second impulse is added to the velocity V at the zero restoring-force timing induced by the first impulse (full recovery of the velocity at the zero restoring-force timing due to zero damping). Since the strain energy does not exist at the zero restoring-force point, the idea in Figure 1.5(a) can be applied to this case.

Consider secondly the case (CASE 2) where the system goes into the yielding stage after the second impulse. Figures 2.1(c) and 2.1(d) show the schematic diagram of the response in this case. As in CASE 1, the maximum deformation $u_{\mathrm{max}1}$ after the first impulse can be obtained from the energy balance law (see Figure 1.5(a)).

$$mV^2/2 = ku_{\mathrm{max}1}^2/2 \qquad (2.4)$$

On the other hand, the maximum deformation $u_{\mathrm{max}2}$ after the second impulse can be computed from another energy balance law (mechanical energy is not conserved in this case).

$$m\left(2V\right)^2/2 = f_y d_y/2 + f_y u_{p2} = f_y d_y/2 + f_y\left(u_{\mathrm{max}2} - d_y\right) \qquad (2.5)$$

As in the above case, the velocity V induced by the second impulse is added to the velocity V at the zero restoring-force timing induced by the first impulse. Please see Figure 1.7 except for the experience of plastic deformation after the first impulse.

Consider finally the case (CASE 3) where the system goes into the yielding stage even after the first impulse. Let us refer to Figures 1.5(b) and 1.7. Figures 2.1(e) and 2.1(f) present the schematic diagram of the response in this case. The maximum deformation $u_{\mathrm{max}1}$ after the first impulse can be obtained from the energy balance law.

$$mV^2/2 = f_y d_y/2 + f_y u_{p1} = f_y d_y/2 + f_y\left(u_{\mathrm{max}1} - d_y\right) \qquad (2.6)$$

On the other hand, the maximum deformation $u_{\mathrm{max}2}$ after the second impulse can be computed from another energy balance law.

$$m\left(v_c + V\right)^2/2 = f_y d_y/2 + f_y u_{p2} \qquad (2.7)$$

where v_c is the velocity of mass at the zero restoring-force point induced by the first impulse and characterized by $mv_c^2/2 = f_y d_y/2$ and $v_c = V_y$. A part of the input kinetic energy provided by the first impulse is dissipated by the plastic deformation and the strain energy corresponding to the yield deformation, which is stored at the maximum deformation point is transformed into the kinetic energy expressed by the velocity v_c at the zero restoring force point in the unloading process. The plastic deformation u_{p2} after the second impulse is characterized graphically by $u_{\mathrm{max}2} + (u_{\mathrm{max}1} - d_y) = d_y + u_{p2}$ ($u_{\mathrm{max}1}$ is the absolute value). In other words, $u_{\mathrm{max}2}$ can be obtained from

$$m\left(v_c + V\right)^2/2 = f_y d_y/2 + f_y\left(u_{\mathrm{max}1} + u_{\mathrm{max}2} - 2d_y\right). \qquad (2.8)$$

As in the above case, the velocity V by the second impulse is added to the velocity v_c (the maximum velocity in the unloading stage) induced by the first impulse.

The maximum deformations for CASEs 1-3 can be summarized as follows:

(CASE 1)

$$u_{max1}/d_y = V/V_y \qquad \text{(Case 1)} \tag{2.9}$$

$$u_{max2}/d_y = 2(V/V_y) \qquad \text{(Case 1)} \tag{2.10}$$

(CASE 2)

$$u_{max1}/d_y = V/V_y \qquad \text{(Case 2)} \tag{2.11}$$

$$u_{max2}/d_y = 0.5\left\{1+(2V/V_y)^2\right\} \qquad \text{(Case 2)} \tag{2.12}$$

(CASE 3)

$$u_{max1}/d_y = 0.5\left\{1+(V/V_y)^2\right\} \qquad \text{(Case 3)} \tag{2.13}$$

$$u_{max2}/d_y = 0.5(3+2V/V_y) \qquad \text{(Case 3)} \tag{2.14}$$

Figure 2.2 presents the schematic diagram of the employed energy balance law. Figure 2.2(a) shows the correspondence of the response of the SDOF model to the critical double impulse with the trajectory in the force-deformation relation. On the other hand, Figure 2.2(b) indicates the energy balance law after the first and second impulses. It can be found that the kinetic energy given by the first impulse is transformed into the strain energy of the SDOF model attaining the maximum displacement. It should be pointed out that, since the critical second impulse giving the maximum displacement after the second impulse is characterized by the point of the zero restoring force, the computation of the strain energy after the second impulse is simple.

Figure 2.3 shows the comparison between the critical timing of the second impulse and the noncritical timing of the second impulse. It can be observed that the critical input of the second impulse leads to the maximum strain energy (maximum deformation after the second impulse). It may be said by analogy that when a person is pushed by the same power in the state of the fastest running speed, the person can attain the furthest point. The proof of the critical timing of the second impulse is shown in Appendix 1.

Figure 2.4 illustrates the maximum deformation $u_{max}/d_y=$ max(u_{max1}/d_y, u_{max2}/d_y) with respect to the input level. Three regions exist corresponding to CASEs 1-3. In CASEs 1 and 2, the maximum deformation u_{max2}/d_y after the second impulse is larger than the maximum deformation u_{max1}/d_y after the

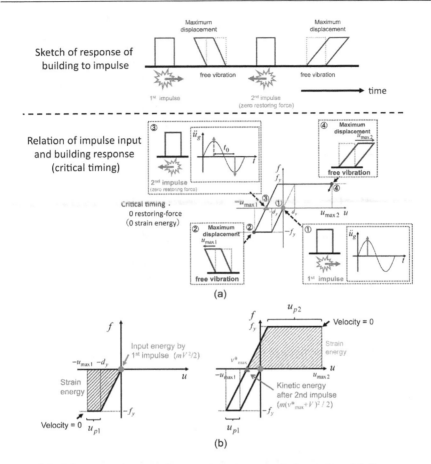

Figure 2.2 Schematic diagram for employed energy balance law, (a) Response of building to critical double impulse, (b) Energy balance law after first and second impulses (kinetic energy is transformed into strain energy).

first impulse. On the other hand, in CASE 3, two regions exist for the boundary case of $u_{max1} = u_{max2}$. While u_{max2}/d_y is larger than u_{max1}/d_y in the smaller input level, u_{max1}/d_y is larger than u_{max2}/d_y in the larger input level. The boundary input level can be calculated by $u_{max1} = u_{max2}$ and is obtained as $V/V_y = 1 + \sqrt{3}$.

Figure 2.5 presents the normalized critical timing t_0/T_1 ($T_1 = 2\pi/\omega_1$) of the second impulse with respect to the input level. In CASEs 1, 2, the SDOF system exhibits an elastic response before the second impulse. Therefore, the critical timing $t_0/T_1 = 0.5$. On the other hand, in CASE 3, the SDOF system shows the elastic-plastic response before the second impulse. The time from the initial state to the input of the second impulse can be expressed as the sum of the time of the initial elastic region, that of the plastic region and that

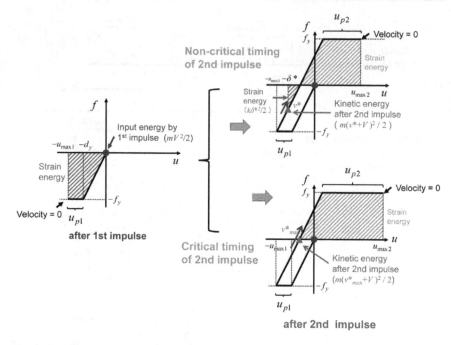

Figure 2.3 Comparison between critical timing of second impulse and noncritical timing of second impulse.

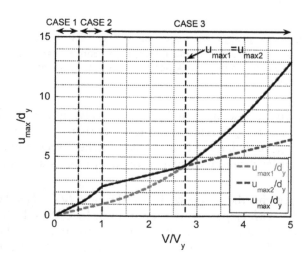

Figure 2.4 Maximum normalized elastic-plastic deformation under double impulse with respect to input level (Kojima and Takewaki 2015).

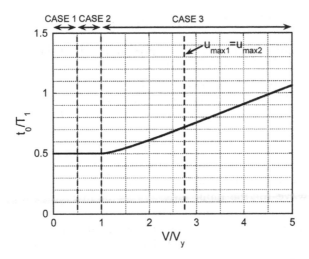

Figure 2.5 Critical interval time of double impulse between first and second impulses with respect to input level (Kojima and Takewaki 2015).

of the unloading region. As stated before, the critical timing of the second impulse coincides with the time corresponding to the zero restoring force in the first unloading process (see Figure 2.1). The passing times of the above three regions can be obtained by solving the differential equation under the initial and terminal conditions. After some manipulation (see Appendix 2), it can be obtained as

$$t_0/T_1 = \left\{\arcsin\left(V_y/V\right)\right\}/(2\pi) + \left\{\sqrt{\left(V/V_y\right)^2 - 1}/(2\pi)\right\} + 1/4 \quad \text{(Case 3)} \quad (2.15)$$

Figure 2.5 shows that the timing is delayed due to plastic deformation as the input level increases. It is important to state again that only critical response giving the maximum value of u_{max2}/d_y is sought by the method explained in this chapter and the critical resonant frequency is obtained automatically for the increasing input level of the double impulse. One of the original points in this chapter is the tracking of only the critical elastic-plastic response.

2.5 ACCURACY INVESTIGATION BY TIME-HISTORY RESPONSE ANALYSIS TO CORRESPONDING ONE-CYCLE SINUSOIDAL INPUT

In order to investigate the accuracy of using the double impulse as a substitute of the corresponding one-cycle sinusoidal wave (representative of the fling-step input), the time-history response analysis of the undamped EPP

SDOF model under the corresponding one-cycle sinusoidal wave was conducted.

In the evaluation procedure, it is important to adjust the input level of the double impulse and the corresponding one-cycle sinusoidal wave based on the equivalence of the maximum Fourier amplitude as noted in Chapter 1 (Section 1.2). The period of the corresponding one-cycle sinusoidal wave is twice that of the time interval of the double impulse $(2t_0)$. Another criterion on the adjustment may be possible from different viewpoints. Figure 2.6 shows one example of the Fourier amplitudes for the input level $V/V_y = 3$. Figures 2.7(a) and (b) illustrate the comparison of the ground displacement and velocity between the double impulse and the corresponding one-cycle sinusoidal wave for the input level $V/V_y = 3$. In Figures 2.6, 2.7(a), (b), $\omega_1 = 2\pi(\text{rad/s})$ $(T_1 = 1.0\text{s})$ and $d_y = 0.16(\text{m})$ are used.

Figure 2.8 presents the comparison of the ductility (maximum normalized deformation) of the undamped EPP SDOF model under the double impulse and the corresponding one-cycle sinusoidal wave with respect to the input level. It can be seen that the double impulse provides a fairly good substitute of the one-cycle sinusoidal wave in the evaluation of the maximum deformation if the maximum Fourier amplitude is adjusted appropriately. Although some discrepancy is observed in the large deformation range $(V/V_y > 3)$, that response range (relatively large plastic deformation) is out of interest in the earthquake structural engineering. The reason for such discrepancy is due to the difference of effect after the first impulse. The modification of the adjustment of the input level between the double impulse and the corresponding one-cycle sine wave can be made in order to guarantee the response correspondence in a broader input range (Chapter 7 and Kojima and Takewaki 2016b).

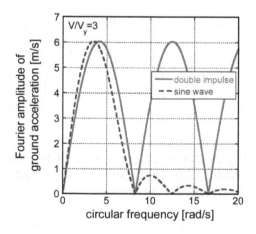

Figure 2.6 Adjustment of input level of double impulse and corresponding one-cycle sinusoidal wave based on Fourier amplitude equivalence (Kojima and Takewaki 2015).

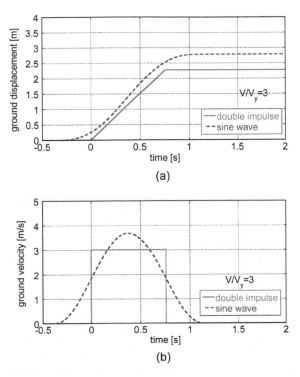

Figure 2.7 Comparison of ground displacement and velocity between double
impulse and corresponding one-cycle sinusoidal wave: (a) Displacement,
(b) Velocity (Kojima and Takewaki 2015).

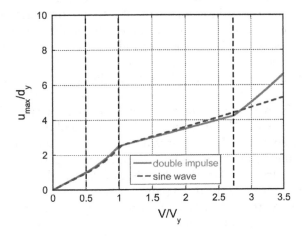

Figure 2.8 Comparison of ductility of elastic-plastic structure under double
impulse and corresponding one-cycle sinusoidal wave (Kojima and
Takewaki 2015).

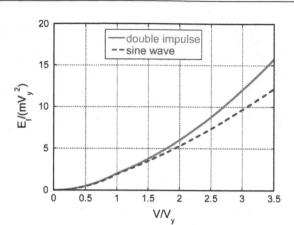

$E_I/(mv_y^2)$

V/V_y

Figure 2.9 Comparison of earthquake input energy by double impulse and corresponding one-cycle sinusoidal wave (Kojima and Takewaki 2015).

Figure 2.9 presents the comparison of the earthquake input energies by the double impulse and the corresponding one-cycle sinusoidal wave. Although a good correspondence can be observed at a lower input level, the double impulse tends to provide a slightly larger value in the larger input level. This property can be understood from the time-history responses shown in Figures 2.10 and 2.11, i.e., a rather clear difference in deformation after the first impulse (the double impulse has a large impact after the first impulse). As stated above, the modification of the adjustment of the input level of the double impulse and the corresponding one-cycle sine wave is possible in order to guarantee the response correspondence in a broader input range (Chapter 7 and Kojima and Takewaki 2016b).

Figure 2.10 illustrates the comparison of response time histories (normalized deformation and restoring-force) under the double impulse and those under the corresponding one-cycle sinusoidal wave. The structural parameters $\omega_1 = 2\pi$(rad/s) $(T_1 = 1.0s)$, $d_y = 0.16$(m) were also used here. While rather good correspondence can be seen in the restoring-force (restoring-force is not sensitive to input variation due to yielding phenomenon), the maximum deformation after the first impulse exhibits a rather larger value in the double impulse. The difference in the initial condition may affect these response discrepancies. However, it is noteworthy that the maximum deformation after the second impulse demonstrates rather good correspondence. This may result from the fact that the effect of the initial condition becomes smaller in this stage and the discrepancy in larger input level in Figure 2.8. Figure 2.11 shows the comparison of the restoring-force characteristic under the double impulse and that under the corresponding one-cycle sinusoidal wave. The structural parameters $\omega_1 = 2\pi$(rad/s) $(T_1 = 1.0s)$, $d_y = 0.16$(m) were also used here. As seen in Figure 2.10, while the maximum deformation after the first impulse exhibits a rather larger value under the double

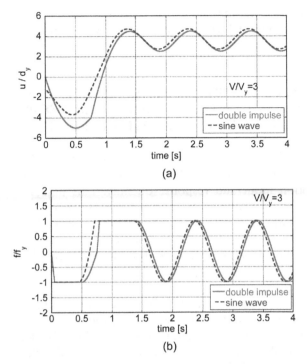

Figure 2.10 Comparison of response time history under double impulse with that under corresponding one-cycle sinusoidal wave: (a) Normalized deformation, (b) Restoring-force (Kojima and Takewaki 2015).

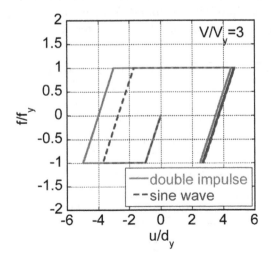

Figure 2.11 Comparison of restoring-force characteristic under double impulse with that under corresponding one-cycle sinusoidal wave (Kojima and Takewaki 2015).

impulse compared to that under the corresponding one-cycle sinusoidal wave, the maximum deformation after the second impulse demonstrates a rather good correspondence.

2.6 DESIGN OF STIFFNESS AND STRENGTH FOR SPECIFIED VELOCITY AND PERIOD OF DOUBLE IMPULSE AND SPECIFIED RESPONSE DUCTILITY

It may be useful to provide a procedure (or flowchart) for the design of stiffness and strength for the specified velocity and period of the near-fault ground motion input and response ductility. This design concept comes from the design philosophy that, if we focus on the worst case of resonance, the safety for other nonresonant cases is guaranteed (see Takewaki 2002).

Nondimensional figures, Figures 2.4 and 2.5, can be used for such design. Figure 2.12 illustrates the procedure (flowchart) for the design of stiffness and strength. One example is presented here:

[Specified conditions] $V=2$(m/s) (velocity of double impulse), $t_0=0.5$(s) (interval of the double impulse and half the period of the corresponding sine wave), $u_{max}/d_y=4.0$ (ductility), $m = 4.0 \times 10^6$(kg)
[Design results] $V/V_y=2.5$, $V_y=0.80$(m/s), $T_1=0.74$(s), $d_y=0.094$(m), $k = 2.9 \times 10^8$(N/m), $f_y = 2.7 \times 10^7$(N)

Figure 2.4 provides $V/V_y=2.5$ for the specified ductility $u_{max}/d_y=4.0$. Then $V_y=0.80$(m/s) is derived from the specified condition $V=2$(m/s) and $V/V_y=2.5$. In the next step, $T_1=0.74$(s) is found from Figure 2.5 for $V/V_y=2.5$ and $t_0=0.5$(s). In this model, $d_y=0.094$(m) is determined from the definition

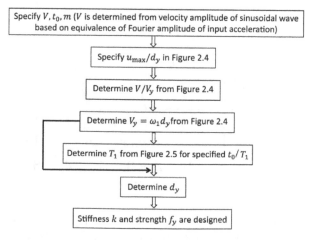

Figure 2.12 Flowchart for design of stiffness and strength.

$V_y = \omega_1 d_y$ and the condition $T_1 (=2\pi/\omega_1)=0.74(s)$. Finally the stiffness $k = 2.9 \times 10^8 (N/m)$ is obtained from $k = \omega_1^2 m$ and the yield strength $f_y = 2.7 \times 10^7 (N)$ is derived from $f_y = kd_y$.

It may be useful to remind that, while most of the previous researches on near-fault ground motions are aimed at disclosing the response characteristics of elastic or elastic-plastic structures with arbitrary stiffness and strength parameters and require tremendous amount of numerical task for clarification, the present chapter focused on the critical response (worst resonant response) and enabled the drastic reduction of computational task. It should be remarked that even the nonresonant case can be treated explicitly by using the energy balance law (see Appendix 1 and Kojima and Takewaki 2016a).

2.7 APPLICATION TO RECORDED GROUND MOTIONS

To demonstrate the applicability of the explained evaluation method, it was applied to two recorded ground motions shown in Chapter 1 (Rinaldi Station FN (Northridge 1994), Kobe Univ. NS (Hyogoken-Nanbu 1995)). Figure 2.13 shows the comparison between the sum of the maximum displacements in both directions and the corresponding value computed by the time-history response analysis. It should be noted that, since the ground motions are specified, the structural parameters, ω_1, d_y, were selected so as for each model to be resonant to the ground motions (refer to Section 7.6 and Kojima and Takewaki 2016a,b for the detailed procedure). It can be observed that the explained method exhibits fairly good performance.

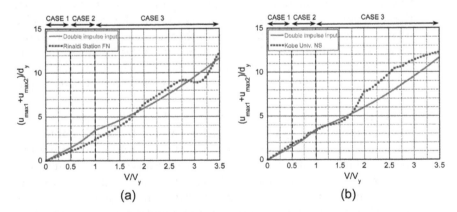

Figure 2.13 Comparison between sum of maximum displacements in both directions to double impulse and corresponding value computed by time-history response analysis to recorded ground motions, (a) Rinaldi Station FN (Northridge 1994), (b) Kobe Univ. NS (Hyogoken-Nanbu 1995) (Kojima and Takewaki 2016a).

2.8 SUMMARIES

The obtained results may be summarized as follows:

1. The double impulse input can be a good substitute of the fling-step (fault-parallel) near-fault ground motion and a closed-form expression can be derived for the maximum response of an undamped elastic-perfectly plastic (EPP) single-degree-of-freedom (SDOF) model under the critical double impulse input.

2. Since only free-vibration is induced under such double impulse input, the energy approach plays an important role in the derivation of the closed-form expression for a complicated elastic-plastic response. In other words, the energy approach enables the derivation of the maximum elastic-plastic seismic response without resorting to solving the differential equation directly. In this process, the input of impulse is expressed by the instantaneous change of velocity of the mass. The maximum elastic-plastic response after impulse can be obtained through the energy balance law by equating the initial kinetic energy computed by the initial velocity to the sum of hysteretic and elastic strain energies. The maximum inelastic deformation can occur either after the first impulse or after the second impulse depending on the input level.

3. The validity and accuracy of the theory explained in this chapter were investigated through the comparison with the response analysis result to the corresponding one-cycle sinusoidal input as a substitute of the main part of the fling-step near-fault ground motion. If the level of the double impulse is adjusted so as for its maximum Fourier amplitude to coincide with that of the corresponding one-cycle sinusoidal wave, the maximum elastic-plastic deformation to the double impulse exhibits a good correspondence with that to the one-cycle sinusoidal wave. The modification of the adjustment of the input level of the double impulse and the corresponding one-cycle sine wave is possible in order to guarantee the response correspondence in a broader input range.

4. While the resonant equivalent frequency has to be computed for a specified input level by changing the excitation frequency parametrically in dealing with the sinusoidal input, no iteration is required in the method for the double impulse. This is because the resonant equivalent frequency can be obtained directly without repetitive procedure. The resonance was proved by using energy investigation and it was made clear that the timing of the second impulse can be characterized by the time corresponding to the zero restoring force in the unloading process.

5. Only critical response (upper bound) was captured by the method and it was shown that the critical resonant frequency can be obtained automatically for the increasing input level of the double impulse. Once the frequency and amplitude of the critical double impulse are computed, the corresponding one-cycle sinusoidal motion can be identified.

6. Using the newly derived nondimensional relations among response ductility, input velocity, and input period, a flowchart was presented for the design of stiffness and strength for the specified velocity and period of the near-fault ground motion input and the specified response ductility. This flowchart can provide a useful result for such design.

REFERENCES

Abbas, A. M., and Manohar, C. S. (2002). Investigations into critical earthquake load models within deterministic and probabilistic frameworks, *Earthquake Eng. Struct. Dyn.*, 31, 813–832.

Caughey, T. K. (1960a). Sinusoidal excitation of a system with bilinear hysteresis, *J. Appl. Mech.*, 27(4), 640–643.

Caughey, T. K. (1960b). Random excitation of a system with bilinear hysteresis, *J. Appl. Mech.*, 27(4), 649–652.

Drenick, R. F. (1970). Model-free design of aseismic structures. *J. Eng. Mech. Div.*, ASCE, 96(EM4), 483–493.

Kojima, K., and Takewaki, I. (2015). Critical earthquake response of elastic-plastic structures under near-fault ground motions (Part 1: Fling-step input), *Frontiers in Built Environment*, 1:12.

Kojima, K., and Takewaki, I. (2016a). Closed-form critical earthquake response of elastic-plastic structures with bilinear hysteresis under near-fault ground motions, *J. of Structural and Construction Eng.* (AIJ), 81(726), 1209–1219 (in Japanese).

Kojima, K., and Takewaki, I. (2016b). Closed-form critical earthquake response of elastic-plastic structures on compliant ground under near-fault ground motions, *Frontiers in Built Environment*, 2:1.

Liu, C.-S. (2000). The steady loops of SDOF perfectly elastoplastic structures under sinusoidal loadings, *J. Marine Science and Technology*, 8(1), 50–60.

Moustafa, A., Ueno, K., and Takewaki, I. (2010). Critical earthquake loads for SDOF inelastic structures considering evolution of seismic waves, *Earthquakes and Structures*, 1(2): 147–162.

Roberts, J. B., and Spanos, P. D. (1990). *Random vibration and statistical linearization*. Wiley, New York.

Shinozuka, M. (1970). Maximum structural response to seismic excitations, *J. Engrg. Mech. Div.*, ASCE, 96(EM5), 729–738.

Takewaki, I. (2002). Robust building stiffness design for variable critical excitations, *J. Struct. Eng.*, ASCE, 128(12), 1565–1574.

Takewaki, I. (2007). *Critical excitation methods in earthquake engineering*, Elsevier, Second edition in 2013.

A. APPENDIX 1: PROOF OF CRITICAL TIMING OF SECOND IMPULSE

Consider the critical timing of the second impulse. Let v_c denote the velocity of the mass passing the point of the zero restoring-force (zero elastic strain energy) in the first unloading process and v^*, u^* denote the velocity of the mass and the elastic deformation component in the first unloading process. Since the first unloading starts from the state with the zero velocity and the elastic strain energy $f_y d_y/2$, the relation $mv_c^2/2 = f_y d_y/2$ holds. From the energy balance law in the unloading process (between the unloading starting point and an arbitrary point in the unloading process), the following relation holds.

$$mv^{*2}/2 + ku^{*2}/2 = f_y d_y/2 \qquad (2.A1)$$

Consider that the second impulse acts at the same time of the state of v^*, u^*. The total mechanical energy just after the action of the second impulse can be expressed by $m(v^* + V)^2/2 + ku^{*2}/2$. By using Eq. (2.A1), the total mechanical energy just after the input of the second impulse can be expressed by

$$m\left(v^* + V\right)^2/2 + ku^{*2}/2$$
$$= mv^{*2}/2 + ku^{*2}/2 + mv^*V + mV^2/2 \qquad (2.A2)$$
$$= f_y d_y/2 + mv^*V + mV^2/2$$

Since the maximum deformation after the second yielding is caused by the power of the maximum total mechanical energy, the maximum velocity among arbitrary velocity v^* causes the maximum deformation after the second yielding. This timing corresponds to the zero restoring-force in the unloading process. This completes the proof.

To ensure the validity of the critical timing shown above, the numerical computation was conducted. Let $t_0{}^c$ denote the critical interval between two impulses. Figure 2.A1 presents the plot of u_{max}/d_y with respect to $t_0/t_0{}^c$. The point of $t_0/t_0{}^c = 1$ indicates the critical timing at zero restoring force. It can be seen that, for larger input velocity level V/V_y, $t_0/t_0{}^c = 1$ is one of the value leading to the maximum value of u_{max}/d_y and, for a smaller level of V/V_y, the point of $t_0/t_0{}^c = 1$ certainly gives the maximum value of u_{max}/d_y. This supports the numerical validation of the proof given above.

B. APPENDIX 2: DERIVATION OF CRITICAL TIMING

In CASE 3 where the SDOF system goes to the plastic range even just after the first impulse, the time from the initial state to the input of the second

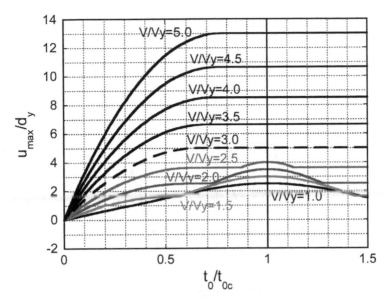

Figure 2.A1 Variation of maximum deformation under double impulse with respect to timing of second impulse (Kojima and Takewaki 2015).

impulse (zero restoring force point) can be expressed as the sum of the time of the initial elastic region, that of the plastic region and that of the unloading region. The times of these three regions can be obtained by solving the differential equation with the substitution of the initial and terminal conditions.

The equation of motion for the initial elastic range can be expressed by

$$m\ddot{x} + kx = 0 \tag{2.A3}$$

In the range of OA in Figure 2.A2, the displacement and velocity can be described by

$$u(t) = -(V/\omega_1)\sin(\omega_1 t) = -(V/V_y)d_y \sin\{2\pi(t/T_1)\} \tag{2.A4}$$

$$\dot{u}(t) = -V\cos(\omega_1 t) = -V\cos\{2\pi(t/T_1)\} \tag{2.A5}$$

The time t_{OA} between point O and A can be characterized by the condition

$$u(t_{OA}) = -(V/\omega_1)\sin(\omega_1 t_{OA}) = -d_y \tag{2.A6}$$

Then the time t_{OA} can be obtained as

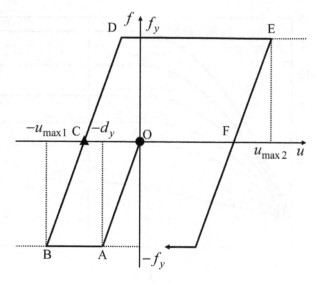

Figure 2.A2 Restoring-force characteristic of SDOF system under double impulse.

$$t_{OA} = \{\arcsin(\omega_1 d_y/V)\}/\omega_1 \qquad (2.A7)$$

The normalized time can be expressed by

$$t_{OA}/T_1 = \{\arcsin(V_y/V)\}/(2\pi) \qquad (2.A8)$$

In this case, the input velocity has to satisfy $0 < \arcsin(V_y/V) < \pi/2$.

Consider next the range AB. The velocity v_A at A can be obtained from the energy balance law.

$$mV^2/2 = (mv_A^2/2) + (kd_y^2/2) \qquad (2.A9)$$

By introducing the expression $\omega_1 d_y = V_y$, v_A can be expressed by

$$v_A = -\sqrt{V^2 - (\omega_1 d_y)^2} = -\sqrt{V^2 - V_y^2} = -\left\{\sqrt{(V/V_y)^2 - 1}\right\}V_y \qquad (2.A10)$$

The equation of motion in the range of AB (plastic range) can be described as

$$m\ddot{x} - f_y = 0 \qquad (2.A11)$$

The velocity and displacement in the range of AB (time is measured from A in the case) are expressed by

$$\dot{x}(t) = \left(\omega_1^2 d_y\right)t + C_1 = V_y\left\{2\pi\left(t/T_1\right)\right\} + C_1 \qquad (2.A12)$$

$$x(t) = \left\{\left(\omega_1^2 d_y\right)/2\right\}t^2 + C_1 t + C_2 = 0.5 d_y\left\{2\pi\left(t/T_1\right)\right\}^2 + C_1 t + C_2 \qquad (2.A13)$$

where C_1 and C_2 are integration constants. The initial conditions $u(t = 0) = - d_y$, $\dot{u}(t = 0) = v_A$ provide the displacement and velocity in the range of AB.

$$
\begin{aligned}
u(t) &= \left\{\left(\omega_1^2 d_y\right)/2\right\}t^2 - \sqrt{V^2 - V_y^2}\,t - d_y \\
&= d_y\left[0.5\left\{2\pi\left(t/T_1\right)\right\}^2 - \left\{\sqrt{\left(V/V_y\right)^2 - 1}\right\}\left\{2\pi\left(t/T_1\right)\right\} - 1\right]
\end{aligned}
\qquad (2.A14)
$$

$$\dot{u}(t) = \left(\omega_1^2 d_y\right)t - \sqrt{V^2 - V_y^2} = V_y\left[\left\{2\pi\left(t/T_1\right)\right\} - \sqrt{\left(V/V_y\right)^2 - 1}\right] \qquad (2.A15)$$

The point B can be characterized by

$$\dot{u}(t = t_{AB}) = \left(\omega_1^2 d_y\right)t_{AB} - \sqrt{V^2 - V_y^2} = 0 \qquad (2.A16)$$

Then the time t_{AB} between A and B can be obtained as

$$t_{AB} = \sqrt{V^2 - V_y^2}/\left(\omega_1^2 d_y\right) \qquad (2.A17)$$

The normalized expression can be expressed by

$$t_{AB}/T_1 = \sqrt{\left(V/V_y\right)^2 - 1}/\left(2\pi\right) \qquad (2.A18)$$

Finally, since the vibration in the process BCD is a free vibration with zero initial velocity, the normalized time in BC can be obtained as

$$t_{BC}/T_1 = 0.25 \qquad (2.A19)$$

The summation of t_{OA}, t_{AB}, t_{BC} is expressed by

$$t_0/T_1 = \left(t_{OA} + t_{AB} + t_{BC}\right)/T_1 = \left[\left\{\arcsin\left(V_y/V\right) + \sqrt{\left(V/V_y\right)^2 - 1}\right\}/\left(2\pi\right)\right] + 0.25 \quad (2.A20)$$

Critical earthquake response of an elastic–perfectly plastic SDOF model under triple impulse as a representative of near-fault ground motions

3.1 INTRODUCTION

In Chapters 1 and 2, the double impulse was introduced as a good representative of near-fault ground motions. In this chapter, a triple impulse is introduced and used as another representative of near-fault ground motions (Kojima and Takewaki 2015b).

Historically, since the science-oriented setting of earthquake ground motion inputs is the basis for the reliable design of structures and infrastructures in the performance-based and resilience-based design, the near-fault ground motions have been investigated extensively and thoroughly from various viewpoints of engineering seismology and earthquake engineering (Bertero et al. 1978, Hall et al. 1995, Sasani and Bertero 2000, Mavroeidis et al. 2004, Alavi and Krawinkler 2004, Bozorgnia and Campbell 2004, Kalkan and Kunnath 2006, 2007, Xu et al. 2007, Rupakhety and Sigbjörnsson 2011, Yamamoto et al. 2011, Vafaei and Eskandari 2015, Khaloo et al. 2015, Kojima et al. 2015, Kojima and Takewaki 2015a). After careful investigations, those ground motions were characterized by the well-known phenomena and concepts called "fling-step" (fault-parallel movement) and "forward-directivity" (fault-normal movement). The detailed explanation of these concepts can be found in the literature (Mavroeidis and Papageorgiou 2003, Bray and Rodriguez-Marek 2004, Kalkan and Kunnath 2006, Mukhopadhyay and Gupta 2013a, b, Zhai et al. 2013, Hayden et al. 2014, Yang and Zhou 2014, Kojima and Takewaki 2015a).

As pointed out in the reference (Kojima and Takewaki 2015a) and Chapter 1 in this book, the fling-step input is a fault-parallel input, and the forward-directivity input is a fault-normal input. The former one is followed by a large deformation slip beneath the ground surface, and the latter one is related to the fault rupture movement and induced perpendicular vibration of the ground. Those ground motions were characterized by two or three wavelets. It was recognized except some cases (Taiwan Chi-Chi in 1999, Kumamoto in 2016, etc.) that the "forward-directivity input" has larger effects on structures in general. Recently, some important investigations were conducted. Mavroeidis and Papageorgiou (2003) explained the characteristics of this class of ground motions in detail and introduced some simple wavelet models (for example, Gabor wavelet and Berlage wavelet). Xu et al. (2007) employed the Berlage wavelet and used it in the performance evaluation of passive control systems. Takewaki and Tsujimoto (2011) made use of the Xu's model and presented a method for scaling ground motions from the viewpoints of story drift and input energy demand (velocity amplitude or acceleration amplitude). Then, Takewaki et al. (2012) employed a sinusoidal wave for pulse-type waves. As a direct extension from a single impulse (Takewaki 2004, Takewaki and Fujita 2009), Kojima and Takewaki (2015a) introduced a simplified input model called "double impulse" (Kojima et al. 2015, Chapter 1 in this book) and derived a closed-form expression for the critical elastic-plastic deformation of a single-degree-of-freedom (SDOF) model to the double impulse input (Chapter 2 in this book). They clarified that (1) a closed-form expression for the critical elastic-plastic deformation can be derived based on a simple energy balance approach and (2) the double impulse can be a good substitute of the fling-step input (one-cycle sinusoidal input) under the equivalence assumption of the maximum Fourier amplitude of accelerations. In this chapter, the approach by Kojima and Takewaki (2015a) explained in Chapter 2 in this book is extended to the forward-directivity input and the intrinsic response characteristics by the forward-directivity are captured (Kojima and Takewaki 2015b).

Most of the former works on the near-fault ground motions focused on the elastic response because there exist a large number of parameters to be considered and the computation of elastic-plastic response itself is quite tedious. In order to resolve the difficulties in such complicated problem, a simple input in terms of the double impulse was employed as a substitute for the fling-step near-fault ground motion recently (Kojima and Takewaki 2015a) and a closed-form expression for the critical elastic-plastic response of a structure by this double impulse was derived (Chapter 2 in this book). The approach will be extended to the forward-directivity input in this chapter (Kojima and Takewaki 2015b). It is shown that, since only free-vibration is induced under such triple impulse input except for the instance of the action of impulses, the energy balance approach plays an important role in the derivation of the closed-form expression for the complicated elastic-plastic response. Different from the situation in the double impulse, a nearly critical excitation is defined, and its response is derived. The word "nearly critical" means that the exact derivation is quite complicated depending on the timing of the second and third impulses. It is also shown that the maximum inelastic deformation can occur either after the second or third impulse depending on the input level. The validity and accuracy of the theory are ensured through the comparison with the time-history response to the corresponding three wavelets of sinusoidal input as a representative of the forward-directivity near-fault ground motion. The amplitude of the triple impulse is modulated so that its maximum Fourier amplitude coincides with that of the corresponding three wavelets of sinusoidal input.

In the history of nonlinear structural dynamics, the closed-form expressions for the elastic-plastic response were obtained only for the steady-state response to an extremely simple sinusoidal input (Caughey 1960, Liu 2000). In the paper (Kojima and Takewaki 2015a, b) and this chapter, the following motivation is posed. If a near-fault ground motion can be represented by double impulse or triple impulse, the critical elastic-plastic response can be derived by an energy approach. As explained for the double impulse, the input of impulse is expressed by the instantaneous change of velocity of the structural mass. The restriction of the response to an almost critical one, which may be interesting in the design stage for safety, enables a unique solution of such complicated elastic-plastic responses.

While the resonant frequency for the equivalent models can be computed for a specified input level by changing the excitation frequency parametrically in the conventional methods dealing with a sinusoidal input (Caughey 1960, Liu 2000), no iteration is required in the method for the triple impulse explained in this chapter. This is because the resonant frequency for the equivalent models (resonance can be proved by using energy investigation) can be obtained directly without the repetitive procedure (the timing of the second impulse can be characterized as the time corresponding to the zero restoring force). In the triple impulse, the analysis can be conducted without the input frequency (timing of impulses) before the second impulse. The criticality is defined only for the response before the input of the third impulse, and it is shown that this restriction is a reasonable condition for the safety evaluation of structures. The critical value of the maximum elastic-plastic response after impulse can be obtained by equating the initial kinetic energy given by the initial velocity to the sum of hysteretic dissipation energy and elastic strain energy. This procedure represents the energy balance law. It should be pointed out that only critical response (upper bound) is captured by the method explained in this chapter, and the critical resonant frequency can be obtained automatically for the increasing input level of the triple impulse.

The significance of using a one-cycle sinusoidal wave and three wavelets of sinusoidal waves as substitutes for fling-step and forward-directivity ground motion inputs were discussed and explained by many researchers (for example, Mavroeidis and Papageorgiou 2003, Kalkan and Kunnath 2006) and the comparison with recorded ground motions was conducted. On the other hand, the merit of the approach explained in the present chapter is to derive a closed-form expression for even elastic-plastic responses under the critical input, which will reduce the computational load drastically and enhance the simplicity of the evaluation of the safety of structures under such near-fault ground motions.

3.2 TRIPLE IMPULSE INPUT

As pointed out in the paper (Kojima and Takewaki 2015a) and Chapter 1 in this book, it is well accepted that the fling-step input (fault-parallel) of the near-fault ground motion can be represented by a one-cycle sinusoidal wave and the forward-directivity input (fault-normal) of the near-fault ground motion can be expressed by three wavelets of sinusoidal input (see Figure 3.1(a), (b)). The latter one is also called "a 1.5-cycle sinusoidal wave" and is similar to the well-known Ricker wavelet (Mavroeidis and Papageorgiou 2003). In this chapter, it is intended to simplify the latter typical near-fault ground motion by the triple impulse. This is because the triple impulse has a simple characteristic, and a straightforward expression on the maximum response can be expected even for elastic-plastic responses based on a simple energy balance approach to free vibrations as in the double impulse in Chapter 2. Furthermore, the triple impulse enables us to describe directly the critical timing of impulses (resonant frequency), which is not possible for the sinusoidal and other inputs without a repetitive procedure.

Consider a ground motion acceleration $\ddot{u}_g(t)$ in terms of the triple impulse, as shown in Figure 3.1(b), expressed by

$$\ddot{u}_g(t) = 0.5V\delta(t) - V\delta(t - t_0) + 0.5V\delta(t - 2t_0) \tag{3.1}$$

where $0.5V$ is the given initial velocity of the first impulse, $\delta(t)$ is the Dirac delta function and t_0 is the time interval among the consecutive two of three impulses. As shown in Figure 3.1(b), the velocity amplitudes $0.5V$, $-V$, $0.5V$ are given so as to attain zero velocity and displacement at the end of input.

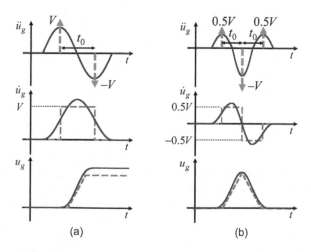

Figure 3.1 Simplification of ground motion (acceleration, velocity, displacement): (a) fling-step input and double impulse; (b) forward-directivity input and triple impulse (Kojima and Takewaki 2015a, b).

The comparison with the corresponding three wavelets of sinusoidal waves as a representative of the forward-directivity input of the near-fault ground motion (Mavroeidis and Papageorgiou 2003, Kalkan and Kunnath 2006) is also plotted in Figure 3.1(b). The corresponding velocity and displacement of such triple impulse and three wavelets of sinusoidal waves are shown in Figure 3.1(b). The Fourier transform of the triple impulse input $\ddot{u}_g(t)$ can be derived as

$$
\begin{aligned}
\ddot{U}_g(\omega) &= \int_{-\infty}^{\infty} \{0.5V\delta(t) - V\delta(t-t_0) + 0.5V\delta(t-2t_0)\} e^{-i\omega t} dt \\
&= \int_{-\infty}^{\infty} \{0.5V\delta(t)e^{-i\omega t} - V\delta(t-t_0)e^{-i\omega t_0}e^{-i\omega(t-t_0)} \\
&\quad + 0.5V\delta(t-2t_0)e^{-i\omega 2t_0}e^{-i\omega(t-2t_0)}\} dt \\
&= V(0.5 - e^{-i\omega t_0} + 0.5e^{-i\omega 2t_0})
\end{aligned} \tag{3.2}
$$

3.3 SDOF SYSTEM

Consider an undamped elastic–perfectly plastic (EPP) single-degree-of-freedom (SDOF) system of mass m and stiffness k. The yield deformation and the yield force are denoted by d_y and f_y, respectively. Let $\omega_1 = \sqrt{k/m}$, u, and f denote the undamped natural circular frequency, the displacement (deformation) of the mass relative to the ground, and the restoring force of the system, respectively. The time derivative is denoted by an over-dot as in Chapter 2.

3.4 MAXIMUM ELASTIC-PLASTIC DEFORMATION OF SDOF SYSTEM TO TRIPLE IMPULSE

The elastic-plastic response of the undamped EPP SDOF system to the triple impulse can be expressed by the continuation of free-vibrations with a sudden change of velocity of mass at the impulse acting points. The maximum deformations after the first, second, and third impulses are denoted by u_{max1}, u_{max2}, and u_{max3}, respectively, as shown in Figure 3.2. The maximum deformations u_{max1}, u_{max2}, and u_{max3} are the absolute value. The input of each impulse results in the instantaneous change of velocity of the structural mass. Such response can be derived by the combination of a simple energy balance approach and the solution of equations of motion for free vibration. It should be remarked that while the solution of equations of motion was not necessary for the double impulse input, this is required in the triple impulse input. This is because a complicated situation exists at the input of the third impulse. The kinetic energy given at the initial stage (the time of action of the first impulse), that at the time of action of the second impulse, and the kinetic energy plus the elastic strain energy at the time of action of the third impulse are transformed into the sum of the hysteretic dissipation

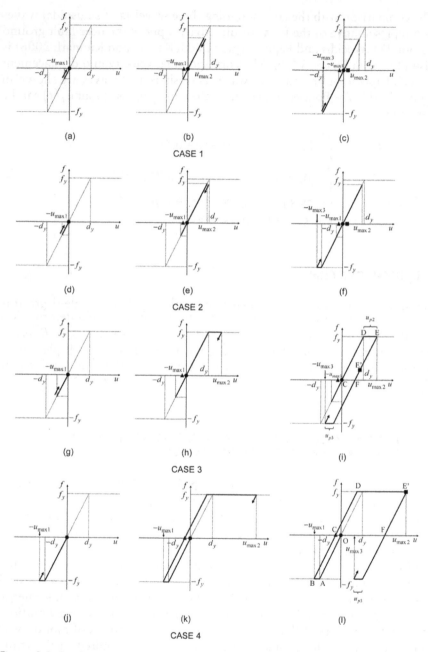

Figure 3.2 Prediction of maximum elastic-plastic deformation under triple impulse based on energy approach: (a) (b) (c) CASE 1: Elastic response, (d) (e) (f) CASE 2: Plastic response after the third impulse, (g) (h) (i) CASE 3: Plastic response after the second impulse, (j) (k) (l) CASE 4: Plastic response after the first impulse (● : first impulse, ▲ : second impulse, ■: third impulse) (Kojima and Takewaki 2015b).

energy and the elastic strain energy corresponding to the yield deformation. By using this rule and incorporating the information from the equations of motion, the maximum deformation can be obtained in a simple manner. It should be noted that while a simple and clear concept of critical input was defined in the case of double impulse (Kojima and Takewaki 2015a and Chapter 2 in this book), the criticality can be used only before the third impulse in the present triple impulse. This is because the timing of action of the third impulse, determined already for the first and second impulses, decreases the maximum deformation u_{max2} after the second impulse and may increase the maximum deformation u_{max3} after the third impulse in the larger input level where the system goes into the yielding stage after the second impulse. However, it is shown that this treatment of determination of timing provides the true criticality in an input level of practical interest.

Although stated in Introduction in this chapter, it should be emphasized again that while the resonant equivalent frequency can be computed for a specified input level by changing the excitation frequency in a parametric or mathematical programming-oriented manner in dealing with the sinusoidal input (Caughey 1960, Liu 2000, Moustafa et al. 2010), no iteration is required in the method for the triple impulse. In the analysis for the triple impulse, the resonant equivalent frequency (resonance can be proved by using energy investigation: see Appendix 1) can be obtained directly without repetition (the timing of the second impulse can be characterized as the time with zero restoring force). It should be reminded again that the resonance is defined before the third impulse.

Only critical response (upper bound for fixed velocity amplitude and variable impulse interval) is captured by the method explained in this chapter, and the critical resonant frequency can be obtained automatically for the increasing input level of the triple impulse. One of the original points in this chapter is the introduction of the concept of "critical excitation" in the elastic-plastic response (Drenick 1970, Abbas and Manohar 2002, Takewaki 2007, Moustafa et al. 2010, Kojima and Takewaki 2015a). Once the frequency and amplitude of the critical triple impulse are computed, the corresponding three wavelets of sinusoidal waves as a representative of the forward-directivity motion can be identified.

Let us explain the procedure for evaluating u_{max1}, u_{max2}, and u_{max3}. Note that the maximum deformations u_{max1}, u_{max2}, and u_{max3} are the absolute value. The plastic deformation after the first impulse is expressed by u_{p1}, that after the second impulse is described by u_{p2}, and that after the third impulse is denoted by u_{p3}. Four cases can be categorized depending on the yielding stage.

CASE 1: Elastic response during all response stages (u_{max3} is the largest)
CASE 2: Yielding after the third impulse (u_{max3} is the largest)
CASE 3: Yielding after the second impulse (u_{max2} or u_{max3} is the largest)

3-1: The third impulse acts in the unloading stage.

3-2: The third impulse acts in the yielding (loading) stage.

CASE 4: Yielding after the first impulse (u_{max2} is the largest)

In comparison with the case of the double impulse, the triple impulse brings difficulties in deriving the critical timing in a general case. This is because the timing of three impulses is fixed by one parameter t_0 and many complicated situations arise. In this chapter, a case is treated where the critical timing is defined only before the input of the third impulse (the critical time interval is obtained from the zero restoring-force timing of the second impulse, after the first impulse). This means that, if the third impulse does not exist, that timing gives the maximum value of u_{max2}. Since the amplitude of the first impulse is half of the second impulse in the triple impulse, it usually occurs that the maximum response after the second impulse is larger than that after the first impulse.

3.4.1 CASE 1

Figures 3.2(a), 3.2(b), and 3.2(c) show the maximum deformations after the first, second, and third impulses, respectively, for the elastic case (CASE 1) during the whole stage. Let us refer to Figure 1.5(a) ($0.5V$ instead of V in this case). The maximum deformation u_{max1} after the first impulse can be obtained from the energy balance law.

$$m(0.5V)^2 /2 = k u_{max1}^2 /2 \tag{3.3}$$

On the other hand, since it can be proved that the critical timing of the second impulse to produce the maximum deformation u_{max2} after the second impulse is the time of zero restoring force (the proof similar to Appendix 1), u_{max2} can be computed from another energy balance law.

$$m(0.5V + V)^2 /2 = k u_{max2}^2 /2 \tag{3.4}$$

In the elastic case, the critical timing of the second impulse is the time of zero restoring force without strain energy, and the velocity V by the second impulse is added to the velocity $0.5V$ induced by the first impulse (full recovery at the zero restoring force due to zero damping). The time interval t_0 is $T_1/2$ in CASE 1. This procedure follows the fact that the input of the impulse causes the sudden change of velocity of the mass.

Furthermore, since the timing of the third impulse is the time of zero restoring force, the maximum deformation u_{max3} after the third impulse can be computed from another energy balance law.

$$m(0.5V + V + 0.5V)^2 /2 = k u_{max3}^2 /2 \tag{3.5}$$

As explained above, the critical timing of the third impulse in the elastic case is the time of zero restoring force, and the velocity $-0.5V$ by the third impulse is added to the velocity $-1.5V$ induced by the first and second impulses (full recovery at the zero restoring force due to zero damping).

It should be remarked again that the critical timing t_0 corresponds to the time of zero restoring force in CASE 1 (see Appendix 1). This is because the time interval between the second and third impulses is equal to that between the first and second impulses in the elastic case. As a result, $u_{\max 3}$ becomes the largest among $u_{\max 1}$, $u_{\max 2}$, and $u_{\max 3}$.

3.4.2 CASE 2

Consider the second case (CASE 2) where the undamped EPP SDOF model goes into the yielding stage after the third impulse. The boundary input velocity level between CASEs 1 and 2 is $V/V_y = 0.5$, and it can be obtained from $u_{\max 3} = d_y$. Here V_y is expressed by $V_y = \omega_1 d_y$ in definition. Figure 3.2(d), 3.2(e), and 3.2(f) show the schematic response for this case. As in CASE 1 (see Figure 1.5(a) ($0.5V$ instead of V)), $u_{\max 1}$ can be obtained from the following energy balance law.

$$m(0.5V)^2/2 = ku_{\max 1}^2/2 \qquad (3.6)$$

On the other hand, $u_{\max 2}$ can be computed from another energy balance law.

$$m(0.5V + V)^2/2 = ku_{\max 2}^2/2 \qquad (3.7)$$

As stated above, the velocity V induced by the second impulse is added to the velocity $0.5V$ at the zero restoring-force point induced by the first impulse. The time interval t_0 is $T_1/2$ in CASE 2. Furthermore, $u_{\max 3}$ can be computed from another different energy balance law.

$$m(0.5V + V + 0.5V)^2/2 = f_y d_y/2 + f_y(u_{\max 3} - d_y) \qquad (3.8)$$

As explained above, the velocity $-0.5V$ induced by the third impulse is added to the velocity $-1.5V$ at the zero restoring-force point induced by the first and second impulses. It should be noted that the critical timing t_0 of the second and third impulses corresponds to the time of zero restoring force, also in CASE 2. This is because the two time intervals coincide due to the elastic response before the third impulse.

3.4.3 CASE 3

Consider the third case (CASE 3) where the model goes into the yielding stage after the second impulse. The boundary input velocity level between

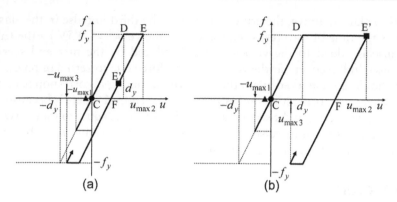

Figure 3.3 Two different cases in CASE 3: (a) CASE3-1, (b) CASE 3-2 (● : first
impulse, ▲ : second impulse, ■: third impulse) (Kojima and Takewaki
2015b).

CASEs 2 and 3 is $V/V_y = 2/3$, and it can be obtained from $u_{max2} = d_y$. Figures
3.2(g), 3.2(h), and 3.2(i) show the schematic response for this case. As in the
previous case, u_{max1} can be obtained from the energy balance law.

$$m(0.5V)^2/2 = ku_{max1}^2/2 \qquad (3.9)$$

As in CASE 2, the critical timing of the second impulse is the time corresponding to the zero restoring force in the unloading process. Although a
more complicated discussion is needed to show the criticality of this timing
depending on the timing of the third impulse, it is omitted here. CASE 3-1
and CASE 3-2 (Figure 3.3) should be considered in CASE 3. In CASE 3-1,
the timing of the third impulse is in the second unloading stage (Figures
3.2(i), 3.3(a)).

3.4.4 CASE 3-1

In this case (CASE 3-1), u_{max2} can be computed from another energy balance
law (as in the same as Eq.(2.5) except the replacement of V by $0.5V$ in the
triple impulse).

$$m(0.5V + V)^2/2 = f_y d_y/2 + f_y(u_{max2} - d_y) \qquad (3.10a)$$

Then, u_{max2} can be expressed as

$$u_{max2} = d_y + m\{(1.5V)^2 - (\omega_1 d_y)^2\}/(2f_y) = 0.5\{(1.5V/V_y)^2 + 1\}d_y \qquad (3.10b)$$

On the other hand, u_{max3} can be derived from another energy balance law.

$$m(v_{E'} - 0.5V)^2/2 + k(\Delta u_{E'F})^2/2 = f_y d_y/2 + f_y u_{p3} \quad (3.11)$$

where $v_{E'}$ (<0) is the velocity at the time of third impulse (point E') and $\Delta u_{E'F} = u_{E'} - u_{p2}$ ($u_{E'}$ (>0): deformation at E'). The plastic deformation u_{p2} after the second impulse is characterized by $u_{p2} = u_{max2} - d_y$, and the plastic deformation u_{p3} after the third impulse satisfies $u_{max2} + u_{max3} = 2d_y + u_{p3}$. The velocity and deformation at point E', $v_{E'}$, and $u_{E'}$ are characterized by Eq. (3.12) and Eq. (3.13), respectively, by solving the equation of motion.

$$v_{E'} = -V_y \sin \omega_1 t_{EE'} \quad (3.12)$$

$$u_{E'} = d_y \cos \omega_1 t_{EE'} + u_{p2} = d_y \cos \omega_1 t_{EE'} + 0.5\left\{(1.5V/V_y)^2 - 1\right\}d_y \quad (3.13)$$

In these equations, $t_{EE'} = (T_1/2) - (t_{CD} + t_{DE})$ is the time interval between the first yielding termination point (point E) and the third impulse acting point (point E'), $t_{CD} = [\arcsin^{-1}\{(V_y/V)(2/3)\}]/\omega_1$ is the time interval between the time of the second impulse (point C) and the time of the first yielding initiation point (point D) (after the second impulse), and $t_{DE} = \left\{\sqrt{(1.5V/V_y)^2 - 1}\right\}/\omega_1$ is the time interval between the time of the first yielding initiation point (point D) (after the second impulse) and the time of the second unloading initiation point (point E). t_{CD} and t_{DE} can be obtained by solving the equations of motion and substituting the transition conditions (yielding and unloading conditions). The time interval t_0 is $T_1/2$ in CASE 3-1. In other words, by paying attention to the direction of the mass motion and the input direction of the third impulse, u_{max3} can be obtained from the energy balance law.

$$m(v_{E'} - 0.5V)^2/2 + k(\Delta u_{E'F})^2/2 = f_y d_y/2 + f_y(u_{max2} + u_{max3} - 2d_y) \quad (3.14a)$$

Then, u_{max3} can be expressed as

$$u_{max3} = d_y + m\left\{(v_{E'}^2 - v_{E'}V - 2V^2) + (\omega_1 \Delta u_{E'F})^2\right\}/(2f_y)$$
$$= d_y + 0.5\left\{(v_{E'}/V_y)^2 - (v_{E'}/V_y)(V/V_y) - 2(V/V_y)^2 + (\Delta u_{E'F}/d_y)^2\right\}d_y \quad (3.14b)$$

If the maximum deformation after the third impulse (corresponding to u_{max3}) is positive, u_{p3} is characterized by $u_{max2} - u_{max3} = 2d_y + u_{p3}$. In this case, u_{max3} can be computed from another energy balance law.

$$m(v_{E'} - 0.5V)^2/2 + k(\Delta u_{E'F})^2/2 = f_y d_y/2 + f_y(u_{max2} - u_{max3} - 2d_y) \quad (3.14c)$$

Then u_{max3} can be expressed by

$$
\begin{aligned}
u_{max3} &= -d_y - m\left\{\left(v_{E'}^2 - v_E.V - 2V^2\right) + \left(\omega_1 \Delta u_{E'F}\right)^2\right\} / \left(2f_y\right) \\
&= -\left[d_y + 0.5\left\{\left(v_{E'}/V_y\right)^2 - \left(v_{E'}/V_y\right)\left(V/V_y\right) - 2\left(V/V_y\right)^2 + \left(\Delta u_{E'F}/d_y\right)^2\right\}d_y\right]
\end{aligned}
\tag{3.14d}
$$

3.4.5 CASE 3-2

In CASE 3-2, the third impulse acts in the yielding stage (Figure 3.3(b)). The boundary input velocity level between CASE 3-1 and CASE 3-2 is $V/V_y = 1.98113\cdots$. It can be obtained numerically by equating $t_{CE}(=t_{CD} + t_{DE})$ to $t_0(=T_1/2)$. In this case (CASE 3-2), u_{max2} indicates the deformation at the time when the third impulse acts (point $u_{E'}$). The deformation $u_{E'}$ can be computed by solving the equation of motion and can be expressed by

$$
u_{E'} = \left\{-0.5\left(\omega_1 t_{DE'}\right)^2 + \sqrt{\left(1.5V/V_y\right)^2 - 1}\left(\omega_1 t_{DE'}\right) + 1\right\}d_y
\tag{3.15}
$$

In this case, u_{max2} is given by

$$
u_{max2} = u_{E'}
\tag{3.16}
$$

On the other hand, u_{max3} can be computed from another energy balance law.

$$
m\left(v_{E'} - 0.5V\right)^2 / 2 + f_y d_y / 2 = f_y d_y / 2 + f_y u_{p3}
\tag{3.17}
$$

In Eq.(3.17), $v_{E'}$ is the velocity at the time when the third impulse acts and u_{p3} is characterized by $u_{max2} - u_{max3} = 2d_y + u_{p3}$. Then $v_{E'}$ is characterized by Eq. (3.18) by solving the equation of motion.

$$
v_{E'} = \left[-\left(\omega_1 t_{DE'}\right) + \sqrt{\left(1.5V/V_y\right)^2 - 1}\right]V_y
\tag{3.18}
$$

In these equations, $t_{DE'} = (T_1/2) - t_{CD}$ is the time interval between the first yielding initiation point (point D in Figure 3.3(b)) and the third impulse acting point (point E' in Figure 3.3(b)), and $t_{CD} = [\arcsin^{-1}\{(V_y/V)(2/3)\}]/\omega_1$ is the time interval between the acting time of the second impulse (zero restoring force) (point C in Figure 3.3(b)) and the time of the first yielding initiation point (point D in Figure 3.3(b)) (after the second impulse). The time t_{CD} is computed by solving the equation of motion and substituting the transition conditions (yielding and unloading conditions). In other words, u_{max3} can be obtained from

$$m\left(v_{E'} - 0.5V\right)^2 / 2 + f_y d_y / 2 = f_y d_y / 2 + f_y \left(u_{\max 2} - u_{\max 3} - 2 d_y\right) \tag{3.19a}$$

Then, $u_{\max 3}$ can be expressed by

$$\begin{aligned} u_{\max 3} &= -2 d_y + u_{E'} - m\left(v_{E'} - 0.5V\right)^2 / \left(2 f_y\right) \\ &= -2 d_y + u_{\max 2} - 0.5\left\{\left(v_{E'} - 0.5V\right) / V_y\right\}^2 \end{aligned} \tag{3.19b}$$

3.4.6 CASE 4

Consider finally the case (CASE 4) where the undamped EPP SDOF model goes into the yielding stage even after the first impulse. The boundary input velocity level between CASEs 3 and 4 is $V/V_y = 2$, and it can be obtained from $u_{\max 1} = d_y$. Let us refer to Figure 1.5(b) (0.5V instead of V in this case). Figures 3.2(j), 3.2(k), and 3.2(l) show the schematic response for this case. As in Figure 1.5(b) except 0.5V in place of V for the double impulse, $u_{\max 1}$ can be obtained from the energy balance law.

$$m\left(0.5V\right)^2 / 2 = f_y d_y / 2 + f_y \left(u_{\max 1} - d_y\right) \tag{3.20}$$

In this case (CASE 4), $u_{\max 2}$ is the deformation at the time when the third impulse acts ($u_{E'}$). The deformation $u_{E'}$ can be computed by solving the equation of motion. Then, $u_{E'}$ can be obtained from Eq. (3.21).

$$\begin{aligned} u_{E'} &= -0.5\left(f_y / m\right) t_{DE'}^2 + \sqrt{V^2 + 2\omega_1 d_y V t_{DE'}} + d_y \\ &\quad - m\left\{\left(V^2 / 4\right) - \left(\omega_1 d_y\right)^2\right\} / \left(2 f_y\right) \\ &= -0.5\left(\omega_1 t_{DE'}\right)^2 d_y + \sqrt{\left(V/V_y\right)^2 + 2\left(V/V_y\right)}\left(\omega_1 t_{DE'}\right) d_y \\ &\quad + 0.5\left\{3 - \left(0.5V/V_y\right)^2\right\} d_y \end{aligned} \tag{3.21}$$

In this case, $u_{\max 2}$ is given by

$$u_{\max 2} = u_{E'} \tag{3.22}$$

On the other hand, by paying attention to the direction of the mass motion and the input direction of the third impulse, $u_{\max 3}$ can be computed from another energy balance law.

$$m\left(v_{E'} - 0.5V\right)^2 / 2 + f_y d_y / 2 = f_y d_y / 2 + f_y u_{p3} \tag{3.23}$$

In Eq.(3.23), $v_{E'}$ is the velocity at the time when the third impulse acts and u_{p3} is characterized by $u_{max2} - u_{max3} = 2d_y + u_{p3}$. Then $v_{E'}$ is characterized by Eq. (3.24) by solving the equation of motion.

$$v_{E'} = -(f_y/m)t_{DE'} + \sqrt{V^2 + 2\omega_1 d_y V}$$
$$= \left[-(\omega_1 t_{DE'}) + \sqrt{(V/V_y)^2 + 2(V/V_y)} \right] V_y \qquad (3.24)$$

In these equations, $t_{DE'} = t_0 - t_{CD} = (t_{OA} + t_{AB} + t_{BC}) - t_{CD}$ is the time interval between the second yielding initiation point (point D in Figure 3.2(l) and the third impulse acting point (point E' in Figure 3.2(l)), and t_0 is the time interval between the first impulse acting point and the second impulse acting point. The time $t_{OA} = \{arcsin^{-1}(2V_y/V)\}/\omega_1$ is the time interval between the first impulse acting point (point O in Figure 3.2(l)) and the first yielding initiation point (point A in Figure 3.2(l)), $t_{AB} = \sqrt{(0.5V/V_y)^2 - 1}/\omega_1$ is the time interval between the first yielding initiation point (point A in Figure 3.2(l)) and the first unloading initiation point (point B in Figure 3.2(l)), $t_{BC} = T_1/4$ is the time interval between the first unloading initiation point (point B in Figure 3.2(l)) and the second impulse acting point (point C in Figure 3.2(l)) and $t_{CD} = [arcsin^{-1}\{V_y/(V_y + V)\}]/\omega_1$ is the time interval between the second impulse acting point (point C in Figure 3.2(l)) and the second yielding initiation point (point D in Figure 3.2(l)). t_{OA}, t_{AB}, t_{BC} and t_{CD} are computed by solving the equations of motion and substituting the transition conditions (yielding and unloading conditions). In other words, u_{max3} can be obtained from

$$m(v_{E'} - 0.5V)^2/2 + f_y d_y/2 = f_y d_y/2 + f_y(u_{max2} - u_{max3} - 2d_y) \quad (3.25a)$$

Then, u_{max3} can be expressed by

$$u_{max3} = -2d_y + u_{E'} - m(v_{E'} - 0.5V)^2/(2f_y)$$
$$= -2d_y + u_{max2} - 0.5\{(v_{E'}/V_y) - 0.5(V/V_y)\}^2 \qquad (3.25b)$$

Figure 3.4 shows the plot of u_{max}/d_y = max $(u_{max1}/d_y, u_{max2}/d_y, u_{max3}/d_y)$ with respect to the input level. $2V_y$ is the input level at which the maximum deformation after the first impulse just attains the yield deformation d_y, and $V/V_y = 2$ is the boundary input velocity level between CASEs 3 and 4. Here V_y is expressed by $V_y = \omega_1 d_y$ in definition. As stated before, there are four cases (CASEs 1-4).

CASE 1: Elastic response during all response stages (u_{max3} is the largest)
CASE 2: Yielding after the third impulse (u_{max3} is the largest)

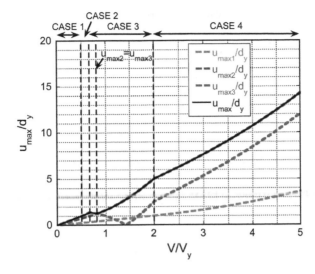

Figure 3.4 Maximum normalized elastic-plastic deformation to triple impulse with respect to input level (Kojima and Takewaki 2015b).

CASE 3: Yielding after the second impulse (u_{max2} or u_{max3} is the largest)
CASE 4: Yielding after the first impulse (u_{max2} is the largest)

The normalized maximum deformation for CASEs 1-4 can be summarized as follows:

$$\frac{u_{max1}}{d_y} = \begin{cases} 0.5\left(V/V_y\right) & \text{for} \quad 0 \le V/V_y < 2.0 \quad (\text{CASEs } 1-3) \\ 0.5\left\{1+\left(0.5V/V_y\right)^2\right\} & \text{for} \quad 2.0 \le V/V_y \quad\quad (\text{CASE } 4) \end{cases} \quad (3.26a)$$

$$\frac{u_{max2}}{d_y} = \begin{cases} 1.5(V/V_y) & \text{for } 0 \le V/V_y < 2/3 \quad (\text{CASEs 1,2}) \\ 0.5\left\{1+(1.5V/V_y)^2\right\} & \text{for } 2/3 \le V/V_y < 1.981 \quad (\text{CASE 3-1}) \\ -0.5(\omega_1 t_{DE'})^2 + \sqrt{(1.5V/V_y)^2 - 1}(\omega_1 t_{DE'}) + 1 & \text{for } 1.981 \le V/V_y < 2.0 \quad (\text{CASE 3-2}) \\ -0.5(\omega_1 t_{DE'})^2 + \sqrt{2(V/V_y) + (V/V_y)^2}(\omega_1 t_{DE'}) & \\ +0.5\left\{3-(0.5V/V_y)^2\right\} & \text{for } 2.0 \le V/V_y \quad (\text{CASE 4}) \end{cases} \quad (3.26b)$$

$$\frac{u_{max3}}{d_y} = \begin{cases} 2(V/V_y) & \text{for } 0 \le V/V_y < 0.5 \quad (\text{CASE 1}) \\ 0.5\left\{1+(2V/V_y)^2\right\} & \text{for } 0.5 \le V/V_y < 2/3 \quad (\text{CASE 2}) \\ \left|0.5(V/V_y)\sin(\omega_1 t_{EE'}) - (V/V_y)^2 + (3/2)\right| & \text{for } 2/3 \le V/V_y < 1.981 \quad (\text{CASE 3-1}) \\ -2 + (u_{max2}/d_y) - 0.5[\left\{-(\omega_1 t_{EE'}) + \sqrt{(1.5V/V_y)^2 - 1}\right\} - 0.5(V/V_y)]^2 & \\ \text{for } 1.981 \le V/V_y < 2.0 \quad (\text{CASE 3-2}) \\ -2 + (u_{max2}/d_y) - 0.5[\left\{-(\omega_1 t_{EE'}) + \sqrt{2(V/V_y) + (V/V_y)^2}\right\} - 0.5(V/V_y)]^2 & \\ \text{for } 2.0 \le V/V_y \quad (\text{CASE 4}) \end{cases} \quad (3.26c)$$

where $t_{EE'}$ and $t_{DE'}$ can be obtained as follows:

$$t_{EE'}/T_1 = 0.5 - \left[\arcsin\left\{ \left(1.5V/V_y \right)^{-1} \right\} + \sqrt{\left(1.5V/V_y \right)^2 - 1} \right] / (2\pi)$$ (3.26d)

for $2/3 \leq V/V_y < 1.981$ (CASE 3-1)

$$\frac{t_{DE'}}{T_1} = \begin{cases} 0.5 - \left\{ 1/(2\pi) \right\} \left[\arcsin\left\{ \left(1.5V/V_y \right)^{-1} \right\} \right] \\ \quad \text{for} \quad 1.981 \leq V/V_y < 2.0 \quad \text{(CASE 3-2)} \\ \left\{ 1/(2\pi) \right\} \left[\arcsin\left\{ 2\left(V_y/V \right) \right\} + \sqrt{\left(0.5V/V_y \right)^2 - 1} \right. \\ \quad \left. - \arcsin\left\{ 1 + \left(V/V_y \right) \right\}^{-1} \right] + 0.25 \\ \quad \text{for} \quad 2.0 \leq V/V_y \quad \text{(CASE 4)} \end{cases}$$ (3.26e)

In both of CASE 1 and 2, u_{max3} is the largest. On the other hand, in CASE 3, u_{max2} or u_{max3} is the largest and in CASE 4, u_{max2} is the largest.

As observed in Figure 3.3, the timing of the third impulse sometimes reduces u_{max2}. It may be useful to assume the timing of the third impulse at the time corresponding to the zero restoring force in the unloading process as shown in Figure 3.5. In both of CASE 1 and 2, this assumption is verified (see Figure 3.2(c), (f)). Figure 3.6 presents the corresponding figure in which the timing of the third impulse is the time corresponding to the zero restoring force after the attainment of u_{max2} (in the process of the second unloading). In Figure 3.6, four cases (CASE 1*, CASE 2*, CASE 3*, CASE 4*) are introduced corresponding to the previously defined four cases. CASE 1* and

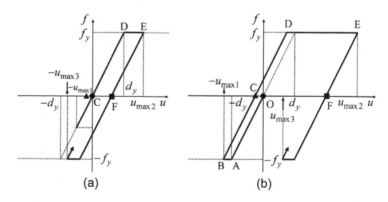

Figure 3.5 Modified version of timing of third impulse: (a) CASE 3*, (b) CASE 4* (● : first impulse, ▲: second impulse, ■: third impulse) (Kojima and Takewaki 2015b).

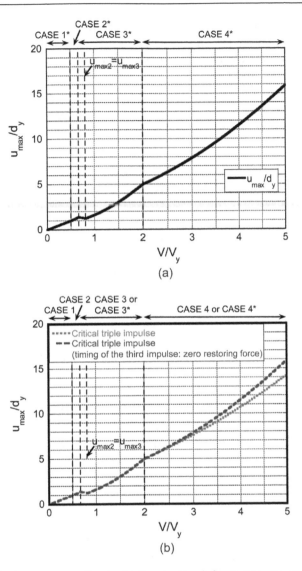

Figure 3.6 Maximum normalized elastic-plastic deformation to triple impulse with respect to input level (timing of the third impulse: zero restoring force): (a) normalized maximum deformation (timing of the third impulse: zero restoring force); (b) comparison of CASEs 3*,4* and CASEs 3, 4 (Kojima and Takewaki 2015b).

CASE 2* are CASE 1 and CASE 2 themselves. It can be realized because the timing of the third impulse defined as the same interval between the first and second impulses decreases u_{max2}, u_{max}/d_y in CASE 4 in Figure 3.4 becomes smaller than that in Figure 3.6. However, it is noteworthy that u_{max}/d_y in Figure 3.6 can be a good upper bound of that in Figure 3.4 and u_{max}/d_y

(u_{max2}/d_y) to any other timing t_0 (except zero restoring force) does not become larger than that in Figure 3.6 (see Appendix 1, 2).

The normalized maximum deformation for CASEs 1, 2, 3* and 4* can be summarized as follows:

$$\frac{u_{max1}}{d_y} = \begin{cases} 0.5(V/V_y) & \text{for } 0 \le V/V_y < 2.0 \quad (\text{CASEs } 1,2,3^*) \\ 0.5\{1+(0.5V/V_y)^2\} & \text{for } 2.0 \le V/V_y \quad\quad (\text{CASE } 4^*) \end{cases} \quad (3.27a)$$

$$\frac{u_{max2}}{d_y} = \begin{cases} 1.5(V/V_y) & \text{for } 0 \le V/V_y < 2/3 \quad (\text{CASE } 1, 2) \\ 0.5\{1+(1.5V/V_y)^2\} & \text{for } 2/3 \le V/V_y < 2.0 \quad (\text{CASE } 3^*) \\ 0.5\{(3/4)(V/V_y)^2 + 2(V/V_y)+3\} & \text{for } 2.0 \le V/V_y \quad\quad (\text{CASE } 4^*) \end{cases} \quad (3.27b)$$

$$\frac{u_{max3}}{d_y} = \begin{cases} 2(V/V_y) & \text{for} \quad 0 \le V/V_y < 0.5 \quad (\text{CASE } 1) \\ 0.5\{1+(2V/V_y)^2\} & \text{for} \quad 0.5 \le V/V_y < 2/3 \quad (\text{CASE } 2) \\ \left| 0.5\{-2(V/V_y)^2 + (V/V_y)+3\} \right| & \\ \quad\quad\quad\quad\quad\quad \text{for} \quad 2/3 \le V/V_y < 2.0 \quad (\text{CASE } 3^*) \\ 0.5\{0.5(V/V_y)^2 + (VEV_y)-1\} & \\ \quad\quad\quad\quad\quad\quad \text{for} \quad 2.0 \le V/V_y \quad (\text{CASE } 4^*) \end{cases} \quad (3.27c)$$

Figure 3.7 presents the normalized interval t_0/T_1 $(T_1 = 2\pi/\omega_1)$ between the first and second impulses with respect to the input level. The critical timing can be obtained as follows.

$$\frac{t_0^c}{T_1} = \begin{cases} 0.5 & \\ \quad \text{for } 0 \le V/V_y < 2.0 \quad (\text{CASEs } 1-3) \\ [\arcsin(0.5V/V_y)^{-1} + \sqrt{(0.5V/V_y)^2 - 1}]/(2\pi)+0.25 \\ \quad \text{for } 2.0 \le V/V_y \quad (\text{CASE } 4) \end{cases} \quad (3.28)$$

As stated before, this timing coincides with the timing corresponding to the zero restoring force after the first unloading (see Figure 3.2). It can be observed that this timing is delayed as the input level increases. It seems noteworthy to state again that only critical response giving the maximum value of u_{max2}/d_y (in case of the timing of the third impulse after the second unloading) is sought by the method in this chapter and the critical resonant frequency is obtained automatically for the increasing input level of the triple impulse. One of the original points in this chapter is the tracking of the "critical" elastic-plastic response.

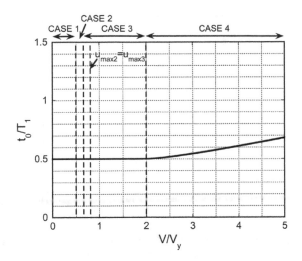

Figure 3.7 Interval time between first and second impulses (the second and third impulses) with respect to input level (Kojima and Takewaki 2015b).

3.5 ACCURACY INVESTIGATION BY TIME-HISTORY RESPONSE ANALYSIS TO CORRESPONDING THREE WAVELETS OF SINUSOIDAL WAVES

To conduct the accuracy investigation of using the triple impulse as a substitute for the corresponding three wavelets of sinusoidal waves (representative of the forward-directivity input), the time-history response analysis of the undamped EPP SDOF model under the three wavelets of sinusoidal waves was conducted.

In the evaluation procedure, it is important to adjust the input level of the triple impulse and the corresponding three wavelets of sinusoidal waves based on the equivalence of the maximum Fourier amplitude. The period of the corresponding three wavelets of sinusoidal waves is twice of the time interval of the triple impulse $(2t_0)$. The adjustment procedure is shown in Appendix 3. Figure 3.8 shows one example of that correspondence for the input level $V/V_y = 3$. Figures 3.9(a) and (b) illustrate the comparison of the ground displacement and velocity between the triple impulse and the corresponding three wavelets of sinusoidal waves for the input level $V/V_y = 3$. Only in Figures 3.8 and 3.9(a), (b) are $\omega_1 = 2\pi(\text{rad/s})$ $(T_1 = 1.0\text{s})$ and $d_y = 0.16(\text{m})$ used.

Figure 3.10 presents the comparison of the ductility (maximum normalized deformation) of the undamped EPP SDOF model under the triple impulse and the corresponding three wavelets of sinusoidal waves with respect to the input level. It can be found that the triple impulse is a good substitute for the three wavelets of sinusoidal waves in the evaluation of the maximum deformation if the maximum Fourier amplitude is adjusted

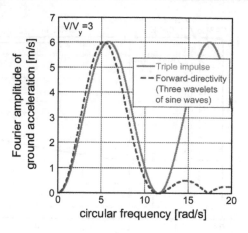

Figure 3.8 Adjustment of input level of triple impulse and corresponding three wavelets of sinusoidal waves based on Fourier amplitude equivalence (Kojima and Takewaki 2015b).

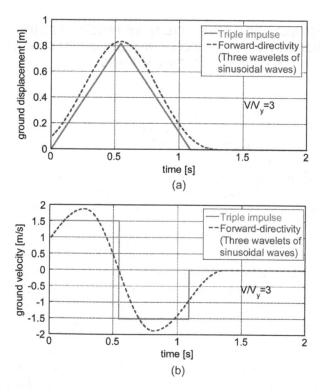

Figure 3.9 Comparison of ground displacement and velocity between triple impulse and corresponding three wavelets of sinusoidal waves: (a) displacement, (b) velocity (Kojima and Takewaki 2015b).

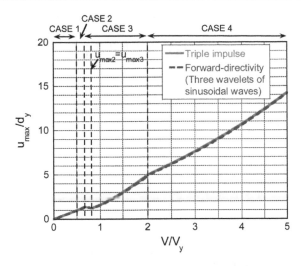

Figure 3.10 Comparison of ductility of elastic-plastic structure to triple impulse and corresponding three wavelets of sinusoidal waves (Kojima and Takewaki 2015b).

appropriately. It should be remarked that, while the correspondence of the maximum deformation between the double impulse and the one-cycle sinusoidal input has some accuracy problem in the large input level ($V/V_y > 3$), the correspondence of the maximum deformation between the triple impulse and the three wavelets of sinusoidal waves is extremely good in a wide range of input level.

Figure 3.11 shows the comparison of the earthquake input energies by the triple impulse and the corresponding three wavelets of sinusoidal waves. An extremely good correspondence can also be observed in the input energy, and this is quite different from the case of the double impulse (the double impulse caused slightly larger input energy compared to the one-cycle sine wave). As stated before, the ratio 0.5 of the amplitude of the first impulse to the second impulse may be the main reason for this good correspondence of the maximum responses. This supports the validity of the triple impulse as a substitute for the forward-directivity near-fault ground motion.

Figure 3.12 illustrates the comparison of response time histories (normalized deformation and restoring-force) under the triple impulse and those under the corresponding three wavelets of sinusoidal waves. The parameters $\omega_1 = 2\pi$(rad/s) ($T_1 = 1.0$s), $d_y = 0.16$(m) are used here. While a rather good correspondence can be seen in general, the amplitude of deformation after the third impulse exhibits a slightly different value resulting from the difference in the applied timing of the third impulse. At the same time, a difference in phase can be observed both in the deformation and restoring-force. This may also result from the difference in the applied timing of the third impulse.

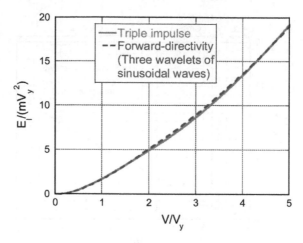

Figure 3.11 Comparison of earthquake input energy by triple impulse and corresponding three wavelets of sinusoidal waves (Kojima and Takewaki 2015b).

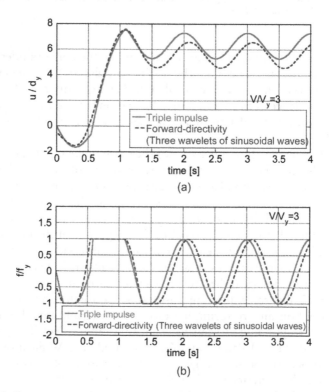

Figure 3.12 Comparison of response time history to triple impulse with that to corresponding three wavelets of sinusoidal waves: (a) normalized deformation, (b) normalized restoring-force (Kojima and Takewaki 2015b).

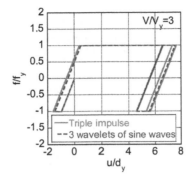

Figure 3.13 Comparison of restoring-force characteristic under triple impulse with that under corresponding three wavelets of sinusoidal waves (Kojima and Takewaki 2015b).

The difference in the amount of energy input at the third impulse seems to influence the later response.

Figure 3.13 presents the comparison of the restoring-force characteristic under the triple impulse and that under the corresponding three wavelets of sinusoidal waves. The parameters $\omega_1 = 2\pi$(rad/s) $(T_1 = 1.0$s), $d_y = 0.16$(m) are also used here. It can be observed from Figure 3.12 that, while the maximum deformations after the first and second impulses exhibit a rather good correspondence, the deformation response after the third impulse exhibits a non-negligible difference. However, since the deformation response after the third impulse does not affect the maximum deformation in an overall time range in a larger input level of CASE 3 and CASE 4, this difference may not be significant.

3.6 DESIGN OF STIFFNESS AND STRENGTH FOR SPECIFIED VELOCITY AND PERIOD OF TRIPLE IMPULSE AND SPECIFIED RESPONSE DUCTILITY

As in the case of the double impulse as a substitute for the near-fault fling-step input, it may be meaningful to present a procedure (flowchart) for the design of stiffness and strength for the specified velocity and period of the near-fault forward-directivity input and the response ductility. This design concept is based on the philosophy that if we focus on the worst resonant case, the safety for other nonresonant cases is guaranteed (see Takewaki 2002). This fact will be explained in the following section.

Since Figures 3.4 and 3.7 are nondimensional ones, they can be used for such design. Figure 3.14 shows the procedure (flowchart) for the design of stiffness and strength. Let us present one example:

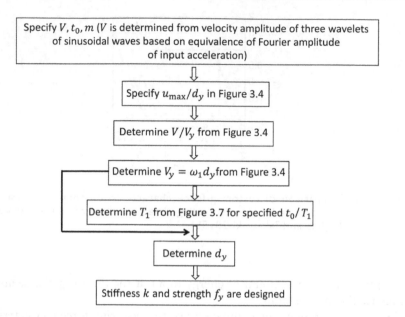

Specify V, t_0, m (V is determined from velocity amplitude of three wavelets of sinusoidal waves based on equivalence of Fourier amplitude of input acceleration)

⇩

Specify u_{max}/d_y in Figure 3.4

⇩

Determine V/V_y from Figure 3.4

⇩

Determine $V_y = \omega_1 d_y$ from Figure 3.4

⇩

Determine T_1 from Figure 3.7 for specified t_0/T_1

⇩

Determine d_y

⇩

Stiffness k and strength f_y are designed

Figure 3.14 Flowchart for design of stiffness and strength (Kojima and Takewaki 2015b).

[Specified conditions] V=2.00(m/s), t_0=0.500(s), u_{max}/d_y=4.00, m = 4.00×
10^6(kg)

[Design results] V/V_y=1.70, V_y=1.18(m/s), T_1=1.00(s), d_y=0.188(m), k=
1.58 × 10^8(N/m), f_y = 2.97 × 10^7(N)

Figure 3.4 provides V/V_y=1.70 for the specified ductility u_{max}/d_y=4.0. Then, V_y=1.18(m/s) is derived from the specified condition V=2.00(m/s) and V/V_y=1.70. In the next step, T_1=1.00(s) is found from Figure 3.7 for V/V_y=1.70 and t_0=0.5(s). In this model, d_y=0.188(m) is determined from the definition $V_y = \omega_1 d_y$ and $T_1(=2\pi/\omega_1)$=1.00(s). Finally the stiffness k = 1.58 × 10^8(N/m) is obtained from $k = \omega_1^2 m$ and the yield force f_y = 2.97 × 10^7(N) is derived from $f_y = kd_y$.

3.7 APPROXIMATE PREDICTION OF RESPONSE DUCTILITY FOR SPECIFIED DESIGN OF STIFFNESS AND STRENGTH AND SPECIFIED VELOCITY AND PERIOD OF TRIPLE IMPULSE

Until Section 3.5, only the critical set of velocity and period of near-fault ground motion input and the corresponding critical response have been treated for a specified design of stiffness and strength. On the other hand, in

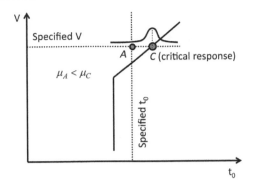

Figure 3.15 Schematic diagram of prediction method of response ductility for specified design of stiffness and strength and specified velocity and period of near-fault ground motion (Kojima and Takewaki 2015b).

Section 3.6, the design flowchart of stiffness and strength for the specified velocity and period of the near-fault ground motion input and specified response ductility was presented. In this section, an approximate method is explained for predicting the response ductility for the specified design of stiffness and strength and the specified velocity and period of near-fault ground motion input. If a more exact response is desired, the time-history response analysis for an arbitrary timing of impulses and an arbitrary input level can be done.

Figure 3.15 shows a schematic diagram for approximately predicting (only predicting the upper bound) the response ductility $\mu = u_{max}/d_y$ for a specified design of stiffness and strength and a specified velocity and period of the near-fault ground motion input using the corresponding critical response. Generally, the specified set of velocity and period of the near-fault ground motion input is not the critical set for a given structure. In such a case, consider the critical set (point C) of velocity and period of the input corresponding to the specified set of velocity. Let μ_A and μ_C denote the response ductilities corresponding to point A and C. From Appendix 1, $\mu_A < \mu_C$ can be shown directly. This enables an approximate prediction of response ductility (only upper bound) for a specified design of stiffness and strength and a specified velocity and period of near-fault ground motion input.

3.8 COMPARISON BETWEEN MAXIMUM RESPONSE TO DOUBLE IMPULSE AND THAT TO TRIPLE IMPULSE

Since it may be meaningful to show the comparison between the maximum response to the double impulse explained in Chapter 2 and that to the triple impulse, it is presented in this section.

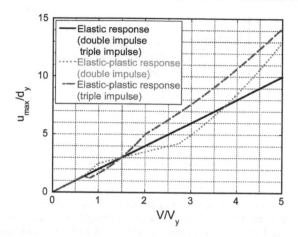

Figure 3.16 Comparison between maximum response to double impulse and that to triple impulse (Kojima et al. 2019).

Figure 3.16 shows such comparison. It can be understood, while the maximum deformation to the double impulse is larger than that to the triple impulse in the range of $0.5 < V/V_y < 1.5$, it reverses in the range of $V/V_y > 1.5$. This phenomenon may be connected to the relation between the magnitudes of the first and second impulses. When the magnitude of the first impulse is the same as the second impulse in the double impulse, the maximum deformation after the second impulse is rather small because of the large deformation in the reverse direction after the first impulse. In the case of the triple impulse, the smaller velocity amplitude $0.5V$ may relax this effect. In Figure 3.16, the maximum deformation of the corresponding elastic model is also plotted for reference. It can be confirmed from the energy balance law that the maximum deformations to the double impulse and the triple impulse are the same.

3.9 APPLICATION TO RECORDED GROUND MOTIONS

To demonstrate the applicability of the explained evaluation method, it was applied to a recorded ground motion shown in Chapter 1 (Rinaldi Station FN during the Northridge earthquake in 1994). Figure 3.17 shows the accelerogram of the Rinaldi station fault-normal component. The main part of this accelerogram is modeled by the 1.5-cycle sinusoidal wave as shown in Figure 3.17. The maximum velocity and the period of the corresponding 1.5-cycle sinusoidal wave are denoted by V_p and T_p. To investigate the effect of variation of the parameter of the sinusoidal wave, $V_p = 1.05[\text{m/s}]$ and $T_p = 0.8, 0.85, 0.9, 0.95[\text{s}]$ were adopted for the Rinaldi station fault-normal component. Although the critical triple impulse was determined for the

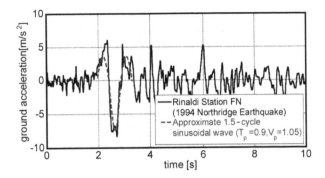

Figure 3.17 Rinaldi station fault-normal component and corresponding 1.5-cycle sinusoidal wave (Kojima et al. 2019).

structural parameter V_y, the structural parameter V_y is selected for the input velocity level of the actual recorded ground motion to maximize the elastic-plastic response under the actual recorded ground motion because the recorded ground motion is fixed (the input velocity level is fixed). This treatment is similar to the case in Section 2.7 (Chapter 2).

Figure 3.18 shows the comparison of the elastic-plastic response under the critical triple impulse and that under the Rinaldi station fault-normal component. The critical elastic-plastic response under the Rinaldi station fault-normal component was obtained by the time-history response analysis. The ordinate axes in Figure 3.18(a) and Figure 3.18(b) present the maximum deformation and the maximum amplitude (positive + negative) of deformation. The abscissa is the normalized input velocity level V/V_y. The input velocity level V of the triple impulse as a substitute for the Rinaldi station fault-normal component is $V = 1.687$[m/s]. Figures 18(a) and (b) demonstrates that the response under the critical triple impulse well simulates the critical elastic-plastic response under the Rinaldi station fault-normal component regardless of some variation of the parameter of the sinusoidal wave.

3.10 SUMMARIES

The obtained results may be summarized as follows:

1. The triple impulse input can be a simplified version of the forward-directivity near-fault ground motion, and a closed-form expression can be derived for the critical response of an undamped elastic-perfectly plastic (EPP) single-degree-of-freedom (SDOF) model under this triple impulse input.

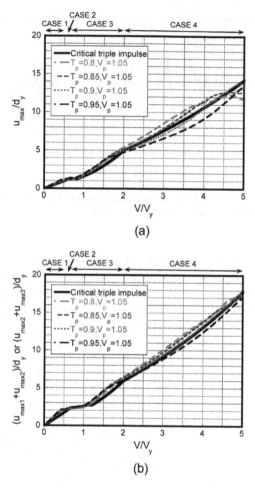

Figure 3.18 Comparison of elastic-plastic response under critical triple impulse and that under Rinaldi station fault-normal component, (a) maximum deformation, (b) maximum deformation amplitude (Kojima et al. 2019).

2. Since only free-vibration is induced under such triple impulse input, the energy approach plays an important role in the derivation of the closed-form expression of a complicated critical elastic-plastic response. In this process, the input of impulse is expressed by the instantaneous change of velocity of the mass. The maximum elastic-plastic response after impulse can be obtained by equating the initial kinetic energy given by the initial velocity to the sum of hysteretic and elastic strain energies. It was shown that the maximum inelastic deformation can occur after either the second or third impulse depending on the input level.

3. The validity and accuracy of the theory explained in this chapter were investigated through the comparison with the time-history response to the corresponding three wavelets of sinusoidal input as a representative of the forward-directivity near-fault ground motion. It was found that if the level of the triple impulse is adjusted so that its maximum Fourier amplitude coincides with that of the corresponding three wavelets of sinusoidal input, the maximum elastic-plastic deformation to the triple impulse exhibits a very good correspondence with that to the three wavelets of sinusoidal waves. This good correspondence is extremely different from the case of the double impulse. The reason may result from the fact that since the velocity level of the second impulse is double of that of the first impulse, the response after the second impulse becomes critical in the broader range of input level in the triple impulse different from the case of the double impulse.

4. While the resonant equivalent frequency has to be computed for a specified input level by changing the excitation frequency parametrically for the sinusoidal input, no iteration is required in the method for the triple impulse. This is because the resonant equivalent frequency (resonance can be proved by using energy investigation) can be obtained directly without resorting to the repetitive procedure (the timing of the second impulse can be characterized as the time corresponding to the zero restoring force). In the triple impulse, the analysis can be conducted without determining the input frequency (timing of impulses) before the second impulse. While a simple and clear concept of critical input was defined in the case of double impulse (Kojima and Takewaki 2015a), the criticality can be used only before the third impulse in the present triple impulse. This is because the timing of the third impulse, determined already for the first and second impulses, decreases the maximum deformation u_{max2} after the second impulse and may increase the maximum deformation u_{max3} after the third impulse.

5. Only critical response (upper bound) is captured by the method, and the critical resonant frequency can be obtained automatically for the increasing input level of the triple impulse. Once the frequency and amplitude of the critical triple impulse are computed, the corresponding three wavelets of sinusoidal motion as a representative of the forward-directivity motion can be identified by the equivalence of the maximum Fourier amplitude.

6. Using the relations among response ductility, input velocity, and input period, a flowchart was presented for the design of stiffness and strength for the specified velocity and period of the near-fault ground motion input and the specified response ductility. It was demonstrated that this flowchart can provide a useful result for such design.

7. An approximate method of prediction of response ductility (only prediction of upper bound) using the corresponding critical response can

be developed for a specified design of stiffness and strength and a specified velocity and period of near-fault ground motion input.

8. The comparison between the maximum response to the double impulse and that to the triple impulse is meaningful. While the maximum deformation to the double impulse is larger than that to the triple impulse in the range of $0.5 < V/V_y < 1.5$, it reverses in the range of $V/V_y > 1.5$. This phenomenon may be connected to the relation between the magnitudes of the first and second impulses.

9. To demonstrate the applicability of the expression for the maximum response to the triple impulse, the maximum response to the critical triple impulse was compared with the critical response under the Rinaldi station FN component during the Northridge earthquake in 1994. The explained method for the triple impulse provides the critical response under the recorded near-fault fault-normal earthquake ground motion with reasonable accuracy.

REFERENCES

Abbas, A. M., and Manohar, C. S. (2002). Investigations into critical earthquake load models within deterministic and probabilistic frameworks, *Earthquake Eng. Struct. Dyn.*, 31, 813–832.

Alavi, B., and Krawinkler, H. (2004). Behaviour of moment resisting frame structures subjected to near-fault ground motions, *Earthquake Eng. Struct. Dyn.*, 33(6), 687–706.

Bertero, V. V., Mahin, S. A., and Herrera, R. A. (1978). Aseismic design implications of near-fault San Fernando earthquake records, *Earthquake Eng. Struct. Dyn.*, 6(1), 31–42.

Bozorgnia, Y., and Campbell, K. W. (2004). Engineering characterization of ground motion, Chapter 5 in *Earthquake engineering from engineering seismology to performance-based engineering*, (eds.) Bozorgnia, Y. and Bertero, V. V., CRC Press, Boca Raton.

Bray, J. D., and Rodriguez-Marek, A. (2004). Characterization of forward-directivity ground motions in the near-fault region, *Soil Dyn. Earthquake Eng.*, 24(11), 815–828.

Caughey, T. K. (1960). Sinusoidal excitation of a system with bilinear hysteresis. *J. Appl. Mech.* 27(4), 640–643.

Drenick, R. F. (1970). Model-free design of aseismic structures, *J. Eng. Mech. Div.*, ASCE, 96(EM4), 483–493.

Hall, J. F., Heaton, T. H., Halling, M. W., and Wald, D. J. (1995). Near-source ground motion and its effects on flexible buildings, *Earthquake Spectra*, 11(4), 569–605.

Hayden, C. P., Bray, J. D., and Abrahamson, N. A. (2014). Selection of near-fault pulse motions, *J. Geotechnical Geoenvironmental Eng.*, ASCE, 140(7).

Kalkan, E., and Kunnath, S. K. (2006). Effects of fling step and forward directivity on seismic response of buildings, *Earthquake Spectra*, 22(2), 367–390.

Kalkan, E., and Kunnath, S. K. (2007). Effective cyclic energy as a measure of seismic demand, *J. Earthquake Eng.*, 11, 725–751.

Khaloo, A. R., Khosravi, H., and Hamidi Jamnani, H. (2015). Nonlinear interstory drift contours for idealized forward directivity pulses using "Modified Fish-Bone" models; *Advances in Structural Eng.*,18(5), 603–627.

Kojima, K., Fujita, K., and Takewaki, I. (2015). Critical double impulse input and bound of earthquake input energy to building structure, *Frontiers in Built Environment*, Volume 1, Article 5.

Kojima, K., and Takewaki, I. (2015a). *Critical earthquake response of elastic-plastic structures under near-fault ground motions (Part 1: Fling-step input)*, Frontiers in Built Environment, Volume 1, Article 12.

Kojima, K., and Takewaki, I. (2015b). Critical earthquake response of elastic-plastic structures under near-fault ground motions (Part 2: Forward-directivity input), *Frontiers in Built Environment*, Volume 1, Article 13.

Kojima, K., Fujita, K., and Takewaki, I. (2019). Double and triple impulses for capturing critical elastic-plastic response properties and robustness of building structures under near-fault ground motions, in '*Resilient Structures and Infrastructures*' edited by Noroozinejad Farsangi, E., Takewaki, I., Yang, T. Y., Astaneh-Asl, A., and Gardoni, P., Springer, pp.225-242.

Liu C.-S. (2000). The steady loops of SDOF perfectly elastoplastic structures under sinusoidal loadings, *J. Marine Science and Technology*, 8(1), 50–60.

Mavroeidis, G. P., and Papageorgiou, A. S. (2003). A mathematical representation of near-fault ground motions, *Bull. Seism. Soc. Am.*, 93(3), 1099–1131.

Mavroeidis, G. P., Dong, G., and Papageorgiou, A. S. (2004). Near-fault ground motions, and the response of elastic and inelastic single-degree-freedom (SDOF) systems, *Earthquake Eng. Struct. Dyn.*, 33, 1023–1049.

Moustafa, A., Ueno, K., and Takewaki, I. (2010). Critical earthquake loads for SDOF inelastic structures considering evolution of seismic waves, *Earthquakes and Structures*; 1(2): 147–162.

Mukhopadhyay, S., and Gupta, V. K. (2013a). Directivity pulses in near-fault ground motions—I: Identification, extraction and modeling, *Soil Dyn. Earthquake Eng.*, 50, 1–15.

Mukhopadhyay, S., and Gupta, V. K. (2013b). Directivity pulses in near-fault ground motions—II: Estimation of pulse parameters, *Soil Dyn. Earthquake Eng.*, 50, 38–52.

Rupakhety, R., and Sigbjörnsson, R. (2011). Can simple pulses adequately represent near-fault ground motions?, *J. Earthquake Eng.* 15,1260–1272.

Sasani, M., and Bertero, V. V. (2000). "Importance of severe pulse-type ground motions in performance-based engineering: historical and critical review," in *Proceedings of the Twelfth World Conference on Earthquake Engineering*, Auckland, New Zealand.

Takewaki, I. (2002). Robust building stiffness design for variable critical excitations, *J. Struct. Eng.*, ASCE, 128(12), 1565 –1574.

Takewaki, I. (2004). Bound of earthquake input energy, *J. Struct. Eng.*, ASCE, 130(9), 1289–1297.

Takewaki, I. (2007). *Critical excitation methods in earthquake engineering*, Elsevier, Second edition in 2013.

Takewaki, I., and Fujita, K. (2009). Earthquake input energy to tall and base-isolated buildings in time and frequency dual domains, *J. of The Structural Design of Tall and Special Buildings*, 18(6), 589–606.

Takewaki, I., and Tsujimoto, H. (2011). Scaling of design earthquake ground motions for tall buildings based on drift and input energy demands, *Earthquakes Struct.*, 2(2), 171–187.

Takewaki, I., Moustafa, A., and Fujita, K. (2012). *Improving the earthquake resilience of buildings: The worst case approach*, Springer, London.

Vafaei, D., and Eskandari, R. (2015). Seismic response of mega buckling-restrained braces subjected to fling-step and forward-directivity near-fault ground motions, *Struct. Design Tall Spec. Build.*, 24, 672–686.

Xu, Z., Agrawal, A. K., He, W.-L., and Tan, P. (2007). Performance of passive energy dissipation systems during near-field ground motion type pulses, *Eng. Struct.*, 29, 224–236.

Yamamoto, K., Fujita, K., and Takewaki, I. (2011). Instantaneous earthquake input energy and sensitivity in base-isolated building, *Struct. Design Tall Spec. Build.*, 20(6), 631–648.

Yang, D., and Zhou, J. (2014). A stochastic model and synthesis for near-fault impulsive ground motions, *Earthquake Eng. Struct. Dyn.*, 44, 243–264.

Zhai, C., Chang, Z., Li, S., Chen, Z.-Q., and Xie, L. (2013). Quantitative identification of near-fault pulse-like ground motions based on energy, *Bull. Seism. Soc. Am.*, 103(5), 2591–2603.

A. APPENDIX I: PROOF OF CRITICAL TIMING

In comparison with the double impulse, the response under the triple impulse is quite complicated, and it seems difficult to derive the critical timing in a general case. This difficulty results from the fact that the timing of three impulses (time interval between the consecutive two of the three impulses) is fixed and there exist many complicated situations. In this chapter, a case is treated where the critical timing is defined only before the third impulse. This means that if the third impulse does not exist, that timing provides the maximum value of u_{max2}. The following explanation is almost the same as in Kojima and Takewaki (2015a) and Chapter 2 in this book for the double impulse.

Consider the critical timing of the second impulse (time interval between the first and second impulses). Let v_c denote the velocity of the mass at the time of passing through the point of zero restoring-force (zero elastic strain energy) in the unloading process and v^*, u^* denote the velocity and the elastic deformation component at an arbitrary point in the unloading process. Since the first unloading starts from the state with the zero velocity and the elastic strain energy $f_y d_y/2$, the relation $mv_c^2/2 = f_y d_y/2$ holds from the energy conservation law. From the energy conservation law in the unloading process (between the first unloading initiation point and an arbitrary point), the following relation holds.

$$mv^{*2}/2 + ku^{*2}/2 = f_y d_y/2 \qquad (3.A1)$$

Consider the second impulse at the same time of the state of v^*, u^*. The total mechanical energy just after the input of the second impulse can be expressed by $m(v^* + V)^2/2 + ku^{*2}/2$. By using Eq.(3.A1), the total mechanical energy just after the input of the second impulse can be reduced to

$$m(v^* + V)^2/2 + ku^{*2}/2$$
$$= mv^{*2}/2 + ku^{*2}/2 + mv^*V + mV^2/2$$
$$= f_y d_y/2 + mv^*V + mV^2/2 \qquad (3.A2)$$

Since the maximum deformation after the second yielding is caused by the maximum total mechanical energy, the maximum velocity v^* in Eq. (3.A2) induces the maximum deformation after the second yielding. This timing is the time of zero restoring-force in the unloading process. This completes the proof.

To ensure the critical time interval of the triple impulse, a numerical computation was conducted. Let t_0^c denote the critical time interval between the consecutive two impulses obtained from Eq.(3.28). Figure 3.A1 presents the plot of u_{max}/d_y with respect to t_0/t_0^c. The point of $t_0/t_0^c = 1$ indicates the critical timing of the second impulse at zero restoring force. It can be seen that, for larger input velocity level ($V/V_y = 2.5, 3.0$ in Figure 3.A1), while t_0^c is not the exact critical time interval of the triple impulse, the proposed approach provides a nearly critical response. On the other hand, for a smaller level of V/V_y ($V/V_y = 1.0, 1.5, 2.0$ in Figure 3.A1), the point of $t_0/t_0^c = 1$ certainly gives the maximum value of u_{max}/d_y.

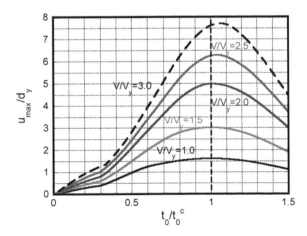

Figure 3.A1 Variation of maximum deformation under triple impulse with respect to timing of second impulse (Kojima et al. 2019).

B. APPENDIX 2: UPPER BOUND OF MAXIMUM RESPONSE VIA RELAXATION OF TIMING OF THIRD IMPULSE

The case is treated here where the third impulse acts at the time corresponding to the zero restoring force in the second unloading process. It can be shown that this case provides the maximum deformation u_{max2} larger than that in the case treated before (the same timing between the first and second impulses).

Consider the case where the model goes into the yielding stage even after the first impulse. The case where the model goes into the yielding stage after the second impulse (elastic after the first impulse) can be explained in almost the same manner. Therefore, that case is not treated here. Figure 3.5(b) shows the schematic response in this case. u_{max1} can be obtained from the energy balance law (refer to Figure 1.5(b) in principle).

$$
\begin{aligned}
m(0.5V)^2/2 &= f_y d_y/2 + f_y u_{p1} \\
&= f_y d_y/2 + f_y \left(u_{max1} - d_y\right)
\end{aligned}
\tag{3.A3}
$$

On the other hand, u_{max2} can be computed from another energy balance law just after the second impulse corresponding to the zero restoring force (refer to Figure 1.7 in principle).

$$
m(v_c + V)^2/2 = f_y d_y/2 + f_y u_{p2}
\tag{3.A4}
$$

where $v_c(=V_y)$ is characterized by $m v_c^2/2 = f_y d_y/2$ and u_{p2} is characterized by $u_{max2} + (u_{max1} - d_y) = d_y + u_{p2}$. In other words, u_{max2} can be obtained from

$$
m(v_c + V)^2/2 = f_y d_y/2 + f_y \left(u_{max1} + u_{max2} - 2d_y\right).
\tag{3.A5}
$$

Furthermore, u_{max3} can be computed from another energy balance law just after the third impulse corresponding to the zero restoring force.

$$
m(v_c + 0.5V)^2/2 = f_y d_y/2 + f_y u_{p3}
\tag{3.A6}
$$

where u_{p3} is characterized by $-u_{max3} + (u_{max2} - d_y) = d_y + u_{p3}$. In other words, u_{max3} can be obtained from

$$
m(v_c + 0.5V)^2/2 = f_y d_y/2 + f_y \left(u_{max2} - u_{max3} - 2d_y\right).
\tag{3.A7}
$$

C. APPENDIX 3: TRIPLE IMPULSE AND CORRESPONDING 1.5-CYCLE SINE WAVE WITH THE SAME FREQUENCY AND SAME MAXIMUM FOURIER AMPLITUDE

The velocity amplitude V of the triple impulse is related to the maximum velocity of the corresponding 1.5-cycle sine wave (three wavelets of sinusoidal waves) with the same frequency so that the maximum Fourier amplitudes of both inputs coincide. The detail is explained here.

The triple impulse is expressed by

$$\ddot{u}_g(t) = 0.5V\delta(t) - V\delta(t - t_0) + 0.5V\delta(t - 2t_0) \qquad (3.A8)$$

where V is the velocity amplitude of the triple impulse, $\delta(t)$ is the Dirac delta function, and t_0 is the time interval between the consecutive two impulses, as shown in Figure 3.1(b). The Fourier transform of Eq. (3.A8) can be obtained as

$$\ddot{U}_g(\omega) = V\left(0.5 - e^{-i\omega t_0} + 0.5e^{-i\omega 2t_0}\right) \qquad (3.A9)$$

Let A_p, T_p, and $\omega_p = 2\pi/T_p$ denote the acceleration amplitude, the period, and the circular frequency of the corresponding 1.5-cycle sine wave (three wavelets of sinusoidal waves), respectively. The acceleration wave $\ddot{u}_g^{SW}$ of the corresponding 1.5-cycle sine wave (Sasani and Bertero 2000, Kalkan and Kunnath 2006, Khaloo et al. 2015) is expressed by

$$\ddot{u}_g^{SW}(t) = \begin{cases} 0.5A_p\sin(\omega_p t) & \left(0 \le t < 0.5T_p, T_p \le t \le 1.5T_p\right) \\ A_p\sin(\omega_p t) & \left(0.5T_p \le t < T_p\right) \end{cases} \qquad (3.A10)$$

The time interval t_0 of two impulses in the triple impulse is related to the period T_p of the corresponding 1.5-cycle sine wave by $T_p = 2t_0$. Although the starting points of both inputs differ by $t_0/2$, the starting time of 1.5-cycle sine wave does not affect the Fourier amplitude. For this reason, the starting time of 1.5-cycle sine wave will be adjusted so that the responses of both inputs correspond well. In this section, the relation of the velocity amplitude V of the triple impulse with the acceleration amplitude A_p of the corresponding 1.5-cycle sine wave is derived. The ratio a_T of A_p to V is introduced by

$$A_p = a_T V \qquad (3.A11)$$

The Fourier transform of $\ddot{u}_g^{SW}$ in Eq.(3.A10) is computed by

$$
\begin{aligned}
\ddot{U}_g^{SW}(\omega) &= \int_0^{t_0} \left\{0.5A_p \sin(\omega_p t)\right\} e^{-i\omega t} dt + \int_{t_0}^{2t_0} \left\{A_p \sin(\omega_p t)\right\} e^{-i\omega t} dt \\
&\quad + \int_{2t_0}^{3t_0} \left\{0.5A_p \sin(\omega_p t)\right\} e^{-i\omega t} dt \\
&= \frac{-0.5A_p \pi t_0}{\pi^2 - (\omega t_0)^2} \left[\begin{array}{l} \{-1 + \cos(\omega t_0) + \cos(2\omega t_0) - \cos(3\omega t_0)\} \\ +i\{-\sin(\omega t_0) - \sin(2\omega t_0) + \sin(3\omega t_0)\} \end{array} \right]
\end{aligned}
\tag{3.A12}
$$

From Eqs. (3.A9) and (3.A12), the Fourier amplitudes of both inputs are expressed by

$$
\left|\ddot{U}_g(\omega)\right| = V\left|\cos(\omega t_0) - 1\right| \tag{3.A13}
$$

$$
\left|\ddot{U}_g^{SW}(\omega)\right| = 0.5A_p \left| \frac{\pi t_0}{\pi^2 - (\omega t_0)^2} \right| \sqrt{ \begin{array}{l} \{-1 + \cos(\omega t_0) + \cos(2\omega t_0) - \cos(3\omega t_0)\}^2 \\ + \{-\sin(\omega t_0) - \sin(2\omega t_0) + \sin(3\omega t_0)\}^2 \end{array} } \tag{3.A14}
$$

It is convenient to shift the time origin by $t = -0.75T_p$. This does not affect the Fourier transform. Then Eq. (3.A10) is transformed into

$$
\ddot{u}_g^{SW}(t) = \begin{cases} -0.5A_p \cos(\omega_p t) & (-0.75T_p \le t < -0.25T_p, \ 0.25T_p \le t < 0.75T_p) \\ -A_p \cos(\omega_p t) & (-0.25T_p \le t < 0.25T_p) \end{cases}
\tag{3.A15}
$$

Fourier transform of Eq. (3.A15) is expressed by

$$
\left|\ddot{U}_g^{SW}(\omega)\right| = A_p \pi t_0 \left|\{\cos(1.5\omega t_0) - \cos(0.5\omega t_0)\} / \{\pi^2 - (\omega t_0)^2\}\right| \tag{3.A16}
$$

The coefficient $a_T(t_0)$ can be derived from Eqs. (3.A11), (3.A13), and (3.A16) and the equivalence of the maximum Fourier amplitude $\left|\ddot{U}_g(\omega)\right|_{max} = \left|\ddot{U}_g^{SW}(\omega)\right|_{max}$.

$$
a_T(t_0) = \frac{A_p}{V} = \frac{\max\left|\cos(\omega t_0) - 1\right|}{\max\left|\pi t_0 \{\cos(1.5\omega t_0) - \cos(0.5\omega t_0)\} / \{\pi^2 - (\omega t_0)^2\}\right|} \tag{3.A17}
$$

In the numerator of Eq. (3.A17), $\max|\cos(\omega t_0) - 1| = 2$. As for denominator, introduce the following function.

$$f_T(x) = \{\cos(1.5x) - \cos(0.5x)\} / (\pi^2 - x^2) \qquad (3.\text{A}18)$$

The maximum value of Eq. (3.A18) is denoted by $f_{T\max}$. After some manipulation, the following relation is derived.

$$a_T(t_0) = 2 / (\pi t_0 f_{T\max}) \qquad (3.\text{A}19)$$

where $f_{T\max}$ = 0.325604587⋯. The integration of $\ddot{u}_g^{SW}(t)$ leads to

$$V_p = A_p / \omega_p \qquad (3.\text{A}20)$$

Eqs. (3.A19) and (3.A20) provide

$$V_p = \{2 / (\pi^2 f_{T\max})\} V \qquad (3.\text{A}21)$$

Then the following relation is derived.

$$V_p / V = 2 / (\pi^2 f_{T\max}) = 0.62235722\ldots \qquad (3.\text{A}22)$$

Eqs. (3.A20) and (3.A22) indicate that V_p/V is constant and A_p/V is a function of t_0 (i.e., $T_p/2$).

Chapter 4

Critical input and response of an elastic–perfectly plastic SDOF model under multi-impulse as a representative of long-duration earthquake ground motions

4.1 INTRODUCTION

In Chapters 2 and 3, the critical excitation methods for near-fault ground motions were explained. In this chapter, the critical excitation method for long-duration ground motions is discussed (Kojima and Takewaki 2015c, Kojima and Takewaki 2017).

There are several types of earthquake ground motions. One is a near-fault ground motion, which is of great interest recently; another is a ground motion of random nature, which is represented by El Centro NS etc.; and still other is a long-duration, long-period ground motion, which was actually observed rather recently and is of special interest and importance for structures of the long natural period (see Takewaki et al. 2011, 2012). As stated in Chapter 1, the effects of near-fault ground motions including typical pulses on the structural response have been investigated extensively (Bertero et al. 1978, Hall et al. 1995, Sasani and Bertero 2000, Mavroeidis et al. 2004, Alavi and Krawinkler 2004, Bozorgnia and Campbell 2004, Kalkan and Kunnath 2006, 2007, Xu et al. 2007, Rupakhety and Sigbjörnsson 2011, Yamamoto et al. 2011, Vafaei and Eskandari 2015, Khaloo et al. 2015). Especially, it was often pointed out that piloti-type buildings with open space in the first story, tall buildings, and base-isolated buildings are greatly affected by such near-fault ground motions. In the field of engineering seismology, the concepts of fling-step and forward-directivity are widely used to characterize such near-fault ground motions (Mavroeidis and Papageorgiou 2003, Bray and Rodriguez-Marek 2004, Bozorgnia and Campbell 2004, Kalkan and Kunnath 2006, Mukhopadhyay and Gupta 2013a, b, Zhai et al. 2013, Hayden et al. 2014, Yang and Zhou 2014). In particular, the Northridge earthquake in 1994, the Hyogoken-Nanbu (Kobe) earthquake in 1995, and the Chi-Chi (Taiwan) earthquake in 1999 drew special attention of many earthquake structural engineers.

As explained in Chapters 1–3, the fling-step (fault-parallel) and forward-directivity (fault-normal) inputs have been characterized by two or three wavelets. The intrinsic natures of those inputs were discussed in Chapters 2 and 3. Many useful investigations have been conducted for this class of ground motions. Following the early studies by Bertero et al. (Bertero et al. 1978, Sasani and Bertero 2000), Mavroeidis and Papageorgiou (2003) investigated the characteristics of this class of ground motions in detail and proposed some simple models (for example, the Gabor wavelet and the Berlage wavelet). Subsequently, Xu et al. (2007) employed a kind of Berlage wavelet and applied it to the performance evaluation of passive energy dissipation systems. Takewaki and Tsujimoto (2011) used the Xu's approach and proposed a method for scaling ground motions from the viewpoints of drift and input energy demand (Takewaki 2004, Takewaki and Fujita 2009). Takewaki et al. (2012) employed a sinusoidal wave as a representative of pulse-type waves.

Most of the former works on the near-fault ground motions deal with the elastic response because the number of parameters (e.g., duration and amplitude of the pulse, ratio of pulse frequency to structure natural frequency, change of equivalent resonant natural frequency for the increased input level) to be considered on this topic is large, and the computation itself of elastic-plastic response is quite complicated. Another reason may be the fact that the elastic displacement response spectra are suitable for clarifying the characteristics of input ground motions.

To remove the difficulties encountered in such important but complicated problem, as a direct extension from a single impulse (Takewaki 2004, Takewaki and Fujita 2009), the double impulse input was introduced by Kojima and Takewaki (2015a) as a substitute of the fling-step near-fault ground motion, and a closed-form expression for the elastic-plastic response of a structure by the "critical double impulse input" was derived (Chapter 2). It was shown that since only free-vibration appears under such double impulse input, the energy balance approach plays an important and critical role in the derivation of the closed-form expression for such a complicated elastic-plastic response. It was also shown that the maximum inelastic deformation can occur either after the first impulse or after the second impulse depending on the input level. The validity and accuracy of the theory were ensured through the comparison with the time-history response to the corresponding one-cycle sinusoidal input as a representative of the fling-step near-fault ground motion. The amplitude of the double impulse was modulated so that its maximum Fourier amplitude coincides with that of the corresponding one-cycle sinusoidal input. The extension of the theory for the fling-step near-fault ground motion to the forward-directivity near-fault ground motion was successfully conducted by Kojima and Takewaki (2015b) (Chapter 3 in this book).

It was pointed out by Takewaki (1996, 1997) that, when considering the upper bound of response to the random earthquake ground motions, it is appropriate to introduce the response spectrum method and the bounding theories (see Takewaki 1996, 1997). In these investigations, an equivalent linearization technique was used to take into account the inelastic response. Therefore, some uncertainties existed in the structural model. On the contrary, the elastic-plastic structural model is used in the method explained in this chapter, as well as in Chapters 2 and 3.

In the long history of nonlinear structural dynamics, the closed-form or nearly closed-form expressions for the elastic-plastic earthquake response were successfully obtained only for the steady-state response to sinusoidal input or the transient response to an extremely simple sinusoidal input (Caughey 1960a, b, Roberts and Spanos 1990, Liu 2000). In this chapter, the following motivation is raised. If a long-duration ground motion can be represented by a multiple impulse, the elastic-plastic response (continuation of free-vibrations) can be derived by an energy balance approach without solving the equation of motion directly. The input of impulse is expressed by the instantaneous change of velocity of the structural mass. The correspondence of the responses to both inputs (multiple impulse and sinusoidal input) can be assured by introducing the equivalence of the maximum Fourier amplitudes of both inputs as in the double and triple impulses. A closed-form expression on the plastic-deformation amplitude is obtained by using an energy balance approach. An approximate expression on the residual displacement is also derived by using the multiple impulse.

In the earthquake-resistant design, the resonance is a keyword for the safety of structures, and it has been investigated extensively. While the resonant equivalent frequency can be computed for a specified input level by changing the excitation frequency in a parametric manner in the conventional methods dealing with the sinusoidal input (Caughey 1960a, b, Iwan 1961, 1965a, b, Roberts and Spanos 1990, Liu 2000), no iteration is required in the method, explained in this chapter, for the multiple impulse (Kojima and Takewaki 2015c). This is supported by the fact that the resonant equivalent frequency can be obtained directly without the repetitive procedure including time-history response analysis. In the multiple impulse, the analysis can be done without the input frequency (timing of impulses) before the second impulse is input. The resonance can be proven by using energy investigation, and the timing of the consecutive impulse can be characterized as the time with zero restoring force. The repetition of this procedure leads to the convergence of the interval time of impulses. The maximum elastic-plastic response after impulse can be obtained by equating the initial kinetic energy given by the initial velocity to the sum of hysteretic and elastic strain energies. It should be pointed out that only critical response (upper bound) is captured by the method explained in this chapter, and the critical resonant frequency can be obtained automatically for the increasing input level of the multiple impulse.

An actual resonant response of a super high-rise building was recorded in Osaka, Japan, during the 2011 off the Pacific coast of Tohoku earthquake. The vibration amplitude (peak-to-peak) of about 3 m was induced at the top as a clear resonant behavior, and slight damage to nonstructural components was reported. This phenomenon clearly indicates the necessity of consideration of response under long-duration ground motion.

4.2 MULTIPLE IMPULSE INPUT

It has been shown by Kalkan and Kunnath (2006) and Khaloo et al. (2015) that the fling-step input (fault-parallel) of the near-fault ground motion can be well represented by a one-cycle sinusoidal wave, and the forward-directivity input (fault-normal) of the near-fault ground motion can be expressed by a series of three sinusoidal wavelets. Recently, Kojima and Takewaki (2015a, b) demonstrated that these sinusoidal waves as the main parts of typical near-fault ground motions can further be modeled by a double impulse (Kojima et al. 2015) and a triple impulse if an appropriate rule of transformation is prepared (Chapters 2–3 in this book). This motivation comes from the fact that the double impulse and triple impulse have a simple characteristic, and a straightforward expression of the maximum response can be expected even for elastic-plastic responses based on an energy balance approach to free vibrations. Although this energy balance approach had been introduced for a single impulse a long time ago (by Thomson 1965), the extension to multiple impulses has never been done. Furthermore, the double and triple impulses enabled us to describe directly the critical timing of impulses (resonant frequency), which is not easy for the sinusoidal and other inputs without a repetitive procedure (Caughey 1960a, b, Iwan 1961, 1965a, b).

Consider a ground acceleration $\ddot{u}_g(t)$ in terms of the multiple impulse, as shown in Figure 4.1(a), expressed by

$$\ddot{u}_g(t) = V\delta(t) - V\delta(t - t_0) + V\delta(t - 2t_0) - V\delta(t - 3t_0) + \cdots \quad (4.1a)$$

where V is the given initial velocity, $\delta(t)$ is the Dirac delta function, and t_0 is the time interval between two consecutive impulses. Its velocity and displacement are shown in Figure 4.1(a). From the realistic point of view in engineering seismology, the following modified multiple input is introduced and treated principally in this chapter (see Figure 4.1(b)).

$$\ddot{u}_g(t) = 0.5V\delta(t) - V\delta(t - t_0) + V\delta(t - 2t_0) - V\delta(t - 3t_0) + \cdots \\ + (-1)^{N-1}V\delta\{t - (N-1)t_0\} + 0.5V\delta(t - Nt_0) \quad (4.1b)$$

where $N + 1$ is the number of impulses, and N is an even number.

The comparison with the corresponding multi-cycle sinusoidal wave as a representative of the long-duration earthquake ground motion is plotted in Figure 4.1(b). The velocity and displacement of both inputs are also plotted in Figure 4.1(b). It can be found that the multiple impulse is a good approximation of the corresponding sinusoidal wave, even in the form of velocity and displacement. Since the responses to similar inputs are often reduced to fairly different ones depending on the response sensitivities, the

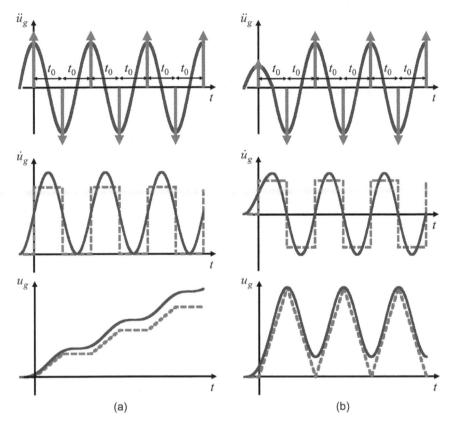

Figure 4.1 Long-duration earthquake ground motion in terms of sinusoidal waves and corresponding multiple impulse: (a) multiple impulse with equal magnitude, (b) multiple impulse with smaller magnitude of first impulse (Kojima and Takewaki 2015c).

correspondence in the response should be discussed carefully. This will be conducted later (Section 4.4, Figure 4.4).

Figure 4.2(a) presents the Input Sequence 1 (original input with equal interval) corresponding to Figure 4.1(b). It can be seen that the points of action of impulses in the force-deformation relation converge to two counterpoints. On the other hand, Figure 4.2(b) shows the Input Sequence 2 (critical timing with residual deformation). The acting points of impulses in the force-deformation relation correspond to the points with zero restoring-force. The time interval between the first and second impulses is found to be different from those between the later consecutive two impulses. However, it may be interesting to note that if we consider the case as shown in Figure 4.2(c), we can reduce the residual displacement to zero by changing the magnitude of the first impulse. Furthermore, if we employ the

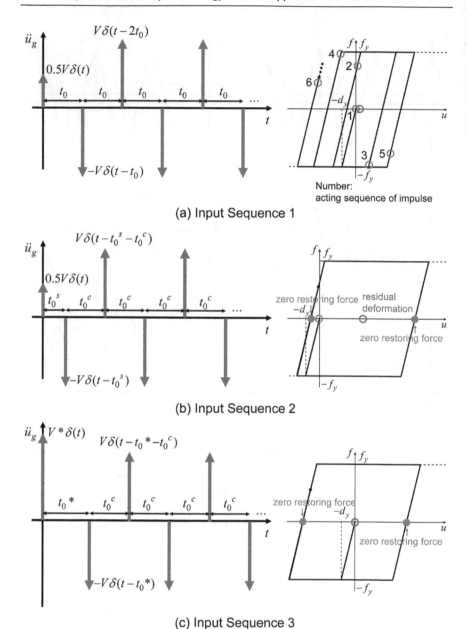

(a) Input Sequence 1

(b) Input Sequence 2

(c) Input Sequence 3

Figure 4.2 Three input sequences: (a) Input Sequence 1: multiple impulse input with equal interval, (b) Input Sequence 2: multiple impulse input with modification of first impulse timing, (c) Input Sequence 3: multiple impulse input with modification of first impulse timing and first impulse amplitude (Kojima and Takewaki 2015c).

critical timing t_0^c in Figure 4.2(b) (Appendix in Kojima and Takewaki (2015a) and Chapter 2) as the timing of the multiple impulse in Figure 4.2(a), the acting points of impulses in the force-deformation relation converge to two counterpoints with zero restoring-force. This fact implies the significance of introducing the Input Sequence 2 in order to find the critical interval of the multiple impulse for the Input Sequence 1 without repetition (the explicit solution of t_0^c is explained in Section 4.4.2). This is the most original aspect of this chapter.

The Fourier transform of $\ddot{u}_g(t)$ of the multiple impulse input (Input Sequence 1 expressed by Eq. (4.1b)) can be expressed as

$$\ddot{U}_g(\omega) = \int_{-\infty}^{\infty} \left\{ \begin{array}{l} 0.5V\delta(t) - V\delta(t-t_0) + V\delta(t-2t_0) - V\delta(t-3t_0) + \cdots \\ +(-1)^{N-1}V\delta\{t-(N-1)t_0\} + 0.5V\delta(t-Nt_0) \end{array} \right\} e^{-i\omega t} dt \qquad (4.2)$$

$$= V\left\{ 0.5 - e^{-i\omega t_0} + e^{-i\omega 2t_0} - e^{-i\omega 3t_0} + \cdots + (-1)^{N-1} e^{-i\omega(N-1)t_0} + 0.5e^{-i\omega Nt_0} \right\}$$

The absolute value of Eq. (4.2) can be expressed by

$$\left| \ddot{U}_g(\omega) \right| = V\left| 0.5 - e^{-i\omega t_0} + e^{-i\omega 2t_0} - e^{-i\omega 3t_0} + \cdots + (-1)^{N-1} e^{-i\omega(N-1)t_0} + 0.5e^{-i\omega Nt_0} \right|$$

$$= V\left| 0.5 + \sum_{n=1}^{N-1} (-1)^n e^{-i\omega nt_0} + 0.5e^{-i\omega Nt_0} \right| \qquad (4.3)$$

4.3 SDOF SYSTEM

Consider an undamped elastic–perfectly plastic (EPP) single-degree-of-freedom (SDOF) system of mass m and stiffness k. The yield deformation and yield force are denoted by d_y and f_y, respectively. Let $\omega_1 = \sqrt{k/m}$, u, and f denote the undamped natural circular frequency, the displacement of the mass relative to the ground (deformation of the system), and the restoring force of the model, respectively. The time derivative is denoted by an overdot as in the previous chapters. In Section 4.4, these parameters will be dealt with in a nondimensional or normalized form to derive the relation of permanent interest between the input and the elastic-plastic response. However numerical parameters will be introduced partially in Sections 4.4 and 4.5.

4.4 MAXIMUM ELASTIC-PLASTIC DEFORMATION OF SDOF SYSTEM TO MULTIPLE IMPULSE

4.4.1 Non-iterative determination of critical timing and critical plastic deformation by using modified input sequence

Firstly, consider Input Sequence 1 as shown in Figure 4.2(a). If the SDOF system is in an elastic range, the critical timing t_0 is exactly a half of the natural period of the SDOF system. On the other hand, if the SDOF system goes into a plastic range, the critical set of input velocity amplitude and input frequency (interval of impulses) need to be computed iteratively. This situation is the same for the multi-cycle sinusoidal wave (Caughey 1960, Iwan 1961, 1965a, b).

To avoid and overcome this difficulty, consider Input Sequence 2, as shown in Figure 4.2(b), which introduces a modified input. In Input Sequence 2, only the time interval between the first and second impulses is modified so that the second impulse is given at the time corresponding to the zero restoring-force. More specifically, Input Sequence 2 was introduced based on the assumption that if the steady state exists in which the impulse is given at the time corresponding to the zero restoring-force, such impulse provides the maximum steady-state plastic deformation. This assumption is assured by giving the critical timing obtained from Input Sequence 2 to Input Sequence 1. In other words, if the critical timing obtained from Input Sequence 2 is given to Input Sequence 1, the timing of impulse converges to the time attaining the zero restoring-force. This fact is also supported by the one-to-one correspondence between the input velocity amplitude and its critical timing of impulses. In this case, impulses have to be given at the times corresponding to the zero restoring-force points. It may also be possible to derive the Input Sequence 3 and its response with zero residual displacement (see Figure 4.2(c)).

The elastic-plastic response of the undamped EPP SDOF model to the multiple impulse can be described by the continuation of free-vibrations after considering sudden change of velocity of mass.

Let u_{max1} and u_{max2} denote the maximum deformations after the first and second impulses in Input Sequence 2, respectively, as shown in Figure 4.3, and let u_p denote the plastic deformation amplitude in the critical steady state, as shown in Figure 4.3. The input of each impulse is expressed by the instantaneous change of velocity of the structural mass. Such responses can be derived by an energy balance approach without resorting to solving the equation of motion directly. The kinetic energy given at the initial stage (the time of action of the first impulse) and at the time of action of the second impulse is transformed into the sum of the hysteretic dissipation energy and the strain energy corresponding to the yield deformation (see Figures 1.5

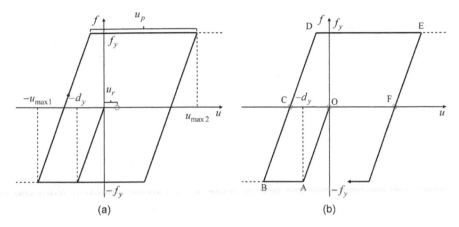

Figure 4.3 Definition of response quantities and response transition: (a) schematic diagram of deformation quantities in force-deformation relation; (b) transition of response process and impulse timing (Input Sequence 2) (Kojima and Takewaki 2015c).

and 1.7). By using this rule, the maximum deformation and plastic deformation amplitude can be obtained in a simple manner.

It should be emphasized that while the resonant equivalent frequency can be computed for a specified input velocity level by moving the excitation frequency parametrically in the conventional methods dealing with the sinusoidal input (Caughey 1960a, b, Iwan 1961, 1965a, b, Roberts and Spanos 1990, Liu 2000, Moustafa et al. 2010), no iteration is required in the method for the multiple impulse explained in this chapter. This is because the resonant equivalent frequency (resonance can be proved by using energy investigation: see Appendix in Kojima and Takewaki 2015a and Chapter 2 in this book) can be obtained directly without the repetitive procedure including nonlinear time-history response analysis. As a result, the timing of the second impulse can be characterized as the time corresponding to the zero restoring force.

Only critical response is captured by the method explained in this chapter, and the critical resonant frequency can be obtained automatically for the increasing input velocity level of the multiple impulse. One of the original points in this chapter is the introduction of the concept of "critical excitation" in the direct elastic-plastic response (Drenick 1970, Abbas and Manohar 2002, Takewaki 2002, 2007, Moustafa et al. 2010). "Direct" means the evaluation without the use of an equivalent linear model for which an iteration is required for determination. Once the frequency and amplitude of the critical multiple impulse are found, the corresponding multi-cycle sinusoidal motion as a representative of the long-duration earthquake ground motion can be identified.

Let us explain the evaluation method of u_{max1}, u_{max2}, and u_p. Let u_{p1} and u_{p2} denote the plastic deformations after the first and second impulses, respectively. In this problem, three cases exist depending on the yielding stage (Kojima and Takewaki 2015a). Let $V_y(=\omega_1 d_y)$ denote the input level of velocity of the impulse at which the undamped EPP SDOF system at rest just attains the yield deformation after the single impulse of such velocity.

Consider the case where the undamped EPP SDOF model goes into the yielding stage even after the first impulse. This case corresponds to CASE 3 in the problem of the double impulse (Kojima and Takewaki 2015a and Chapter 2) (the input level of the first impulse is $0.5V$ instead of V). Figure 4.3(a) shows the schematic diagram of the response in this case. The maximum deformation after the first impulse u_{max1} can be obtained from the following energy balance law (see Eq. (3.20) for the triple impulse).

$$m(0.5V)^2/2 = f_y d_y/2 + f_y u_{p1} = f_y d_y/2 + f_y (u_{max1} - d_y) \qquad (4.4)$$

On the other hand, the maximum deformation after the second impulse u_{max2} can be computed from another energy balance law (see Figure 2.2(b) for the double impulse).

$$m(v_c + V)^2/2 = f_y d_y/2 + f_y u_{p2} \qquad (4.5a)$$

where $v_c(=V_y)$ is characterized by $mv_c^2/2 = f_y d_y/2$ and u_{p2} is characterized by $u_{max2} + (u_{max1} - d_y) = d_y + u_{p2}$. By using this relation, u_{max2} can be obtained from

$$m(v_c + V)^2/2 = f_y d_y/2 + f_y (u_{max1} + u_{max2} - 2d_y). \qquad (4.5b)$$

From Eqs. (4.5a) and (4.5b), $u_p(=u_{p2})$ and u_{max2} can be obtained as follows.

$$u_p/d_y = u_{p2}/d_y = 0.5\left\{(V/V_y)^2 + 2(V/V_y)\right\}, \qquad (4.6a)$$

$$u_{max2}/d_y = 0.5\left\{(3/4)(V/V_y)^2 + 2(V/V_y) + 3\right\}. \qquad (4.6b)$$

As in the above case, the velocity V induced by the second impulse is added to the velocity v_c introduced by the first impulse. This velocity v_c indicates the maximum velocity in the unloading process. Although only CASE 3 in the double impulse (Kojima and Takewaki 2015a and Chapter 2) was considered here, CASE 2 (yielding only after the second impulse) can be treated by replacing v_c in Eqs. (4.5a, b) by $0.5V$. It is noted that when the input velocity V is smaller than 2, the number of impulses more than two is

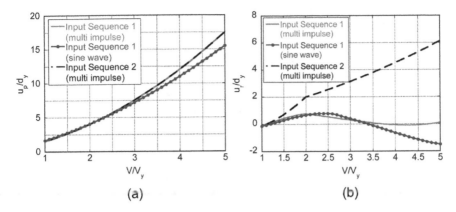

Figure 4.4 Comparison of response among Input Sequence I (multiple impulse), Input Sequence I (sine wave) and Input Sequence 2 (multiple impulse): (a) plastic deformation amplitude; (b) residual deformation (Kojima and Takewaki 2015c).

necessary to attain the inelastic critical steady state in Input Sequence 2, but the plastic deformation amplitude can be calculated by Eq. (4.6a) in all input level.

Figure 4.4 shows the plastic deformation amplitude u_p (u_{p2} in this case) and the residual deformation u_r, shown in Figure 4.3(a), with respect to the input level V/V_y for Input Sequence 1 and 2. It is interesting to note that while the plastic deformation amplitude is the same for Sequence 1 and 2, the residual deformations are different. This results from the difference in the initial disturbances in Input Sequence 1 and 2.

4.4.2 Determination of critical timing of impulses

Consider the Input Sequence 2 (see Figure 4.2) in this section. The time interval between two consecutive impulses can be obtained by solving the equations of motion and substituting the continuation conditions at the transition points. The time interval $t_0{}^s$ between the first and second impulses and the time interval $t_0{}^c$ between two consecutive impulses after the second impulse can be expressed by

$$t_0{}^s/T_1 = (t_{OA} + t_{AB} + t_{BC})/T_1 \qquad (4.7a)$$

$$t_0{}^c/T_1 = (t_{CD} + t_{DE} + t_{EF})/T_1 \qquad (4.7b)$$

where t_{OA}, t_{AB}, t_{BC}, t_{CD}, t_{DE}, and t_{EF} are the time intervals between two consecutive transition points shown in Figure 4.3(b). If $V/V_y < 2$ is satisfied, $t_0{}^s/T_1 = 1/2$ holds. On the other hand, $t_0{}^c$ is obtained from Eq. (4.7b) in all input velocity level. The time intervals t_{OA}, t_{AB}, t_{BC}, t_{CD}, t_{DE}, and t_{EF} are expressed by

$$t_{OA}/T_1 = \left\{\arcsin\left(2/\bar{V}\right)\right\}/(2\pi) \tag{4.8a}$$

$$t_{AB}/T_1 = \sqrt{\left(\bar{V}/2\right)^2 - 1}/(2\pi) \tag{4.8b}$$

$$t_{BC}/T_1 = 1/4 \tag{4.8c}$$

$$t_{CD}/T_1 = \left\{\arcsin\left(1/\left(1+\bar{V}\right)\right)\right\}/(2\pi) \tag{4.8d}$$

$$t_{DE}/T_1 = \sqrt{\left(\bar{V}\right)^2 + 2\bar{V}}/(2\pi) \tag{4.8e}$$

$$t_{EF}/T_1 = 1/4 \tag{4.8f}$$

In Eqs. (4.8a–f), $\bar{V}$ denotes the normalized velocity input V/V_y.

Figures 4.5(a), (b) show two normalized time intervals $t_0{}^s/T_1$ and $t_0{}^c/T_1$ with respect to the input velocity level V/V_y. These time intervals coincide with the time intervals between the points corresponding to the zero restoring force (see Figure 4.3). The timing is delayed due to the existence of plastic deformation as the input level increases. Only the critical response giving the maximum value of u_p/d_y is sought by the method explained in this chapter, and the critical resonant frequency is obtained automatically without repetition for the increasing input level of the multiple impulse. This idea is almost the same as in the case of the double and triple impulses. One of the original points in this chapter is the tracking of only the critical elastic-plastic response.

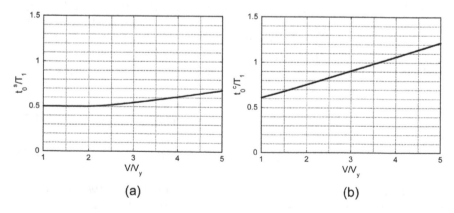

Figure 4.5 Normalized timing $t_0{}^s/T_1$ and $t_0{}^c/T_1$ with respect to input level: (a) $t_0{}^s/T_1$, (b) $t_0{}^c/T_1$ (Kojima and Takewaki 2015c).

4.4.3 Correspondence of responses between input sequence 1 (original one) and input sequence 2 (modified one)

Figures 4.6 and 4.7 present the time-histories of relative displacement (relative to the base motion), relative velocity and restoring-force, and the force-deformation relation under Input Sequence 1 with the resonant impulse interval $t_0/t_0^c = 1.0$ for the input velocity levels $V/V_y = 2$ and $V/V_y = 3$, respectively. In Figures 4.6 and 4.7, the structural parameters $\omega_1 = 2\pi$[rad/sec] $(T_1 = 1.0$[sec]) and $d_y = 0.16$[m] are used. Since the steady state is very sensitive to the time increment in the time-history response analysis using an EPP model, the time increment was chosen as 1.0×10^{-6}[sec]. This phenomenon is well known as plastic flow. In fact, an EPP model was not treated in most works (Caughey 1960a, b, Iwan 1961, 1965a, b) for its difficulty in the treatment of unstable and input-sensitive states. It should be remarked that the impulse timing is the critical one obtained by using the Input Sequence 2. The circles in Figures 4.6 and 4.7 present the acting points of impulses. It

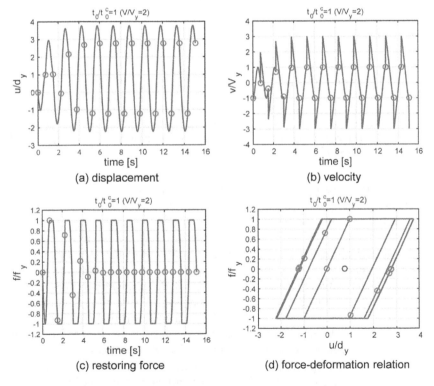

Figure 4.6 Response to Input Sequence 1 for $V/V_y = 2$ (impulse timing is critical one obtained by using Input Sequence 2): (a) displacement, (b) velocity, (c) restoring force, (d) force-deformation relation (Kojima and Takewaki 2015c).

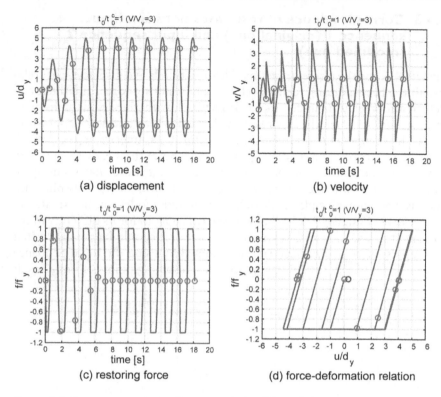

Figure 4.7 Response to Input Sequence I for V/V_y = 3 (impulse timing is critical one obtained by using Input Sequence 2): (a) displacement, (b) velocity, (c) restoring force, (d) force-deformation relation (Kojima and Takewaki 2015c).

can be seen that, although some irregularities appear at first, the response converges gradually to a state with the acting timing of impulse at the zero restoring-force point irrespective of the input velocity level.

Figure 4.8 illustrates the force-deformation relation under the multiple impulse of Input Sequence 1 with three time intervals t_0/t_0^c = 0.8, 1.0, 1.2 for two input levels V/V_y = 2, 3. While, in Figures 4.6 and 4.7, only the case of t_0/t_0^c = 1.0 was treated, three cases t_0/t_0^c = 0.8, 1.0, 1.2 of impulse time intervals are investigated in Figure 4.8. It can be assured that the response converges to a steady state irrespective of the impulse timing and the time interval t_0/t_0^c = 1.0 certainly provides the maximum plastic deformation amplitude u_p. Furthermore, the time interval t_0/t_0^c = 1.0 converges to the applied times of multiple impulses corresponding to the zero restoring force (both directions at counter locations). This indicates the validity of introducing the Input Sequence 2 for finding the critical timing of multiple impulse even for Input Sequence 1.

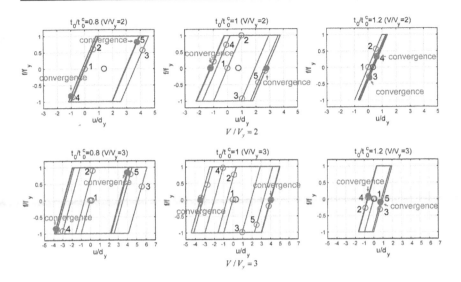

Figure 4.8 Response to Input Sequence I for three impulse timings $(t_0/t_0^c = 0.8,$ 1.0, 1.2) with two input levels V/V_y=2, 3 (Number beside circle: acting sequence of impulse) (Kojima and Takewaki 2015c).

On the other hand, Figures 4.9 and 4.10 present the time-histories of relative displacement to the base motion, relative velocity and restoring-force together with the force-deformation relation under Input Sequence 2 for the input velocity levels $V/V_y = 2$ and $V/V_y = 3$, respectively. In Figures 4.9 and 4.10, the structural parameters $\omega_1 = 2\pi[\text{rad/sec}]$ $(T_1 = 1.0[\text{sec}])$ and $d_y = 0.16[\text{m}]$ were used. The response exhibits a steady state from the initial stage and corresponds to a state with the timing of impulse at the zero restoring-force point irrespective of the input velocity level. It is seen that the critical timing t_0^c of multiple impulse computed by Eq. (4.7b) can be obtained without repetition and can be used as the critical timing even for the Input Sequence 1.

4.5 ACCURACY INVESTIGATION BY TIME-HISTORY RESPONSE ANALYSIS TO CORRESPONDING MULTI-CYCLE SINUSOIDAL INPUT

To check the accuracy in using the multiple impulse (Input Sequence 1) as a substituted version of the corresponding multi-cycle sinusoidal wave (representative of the long-duration ground motion input), the time-history response analysis of the undamped EPP SDOF model under the multi-cycle sinusoidal wave was conducted.

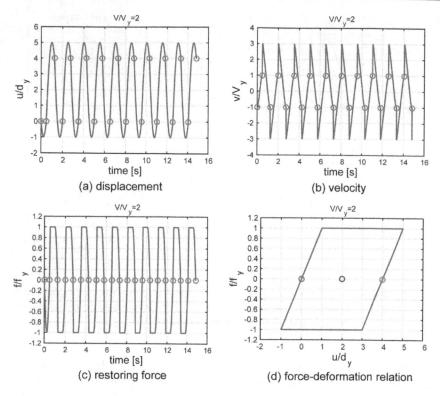

Figure 4.9 Response to Input Sequence 2 for $V/V_y = 2$: (a) displacement, (b) velocity, (c) restoring force, (d) force-deformation relation (Kojima and Takewaki 2015c)

In this investigation, it is important to adjust the input level of the multiple impulse and the corresponding multi-cycle sinusoidal wave. This adjustment is conducted by using the equivalence criterion of the maximum Fourier amplitude (see Appendix 1) and a modification based on the response equivalence at some points with different input levels for better correspondence in the wider input velocity range (point fitting processing).

Figures 4.11(a) and (b) show the comparison of the ground displacement and velocity between the multiple impulse and the corresponding multi-cycle sinusoidal wave for the input level $V/V_y = 3$. In Figures 4.11(a) and (b), the structural parameters $\omega_1 = 2\pi$(rad/s) ($T_1 = 1.0$s) and $d_y = 0.16$(m) were used. The amplitude of the sinusoidal wave was amplified by 1.15 after both Fourier amplitudes of the sinusoidal wave, and the multiple impulse were adjusted (10 cycles). This amplification factor 1.15 was introduced based on the response equivalence at some points with different input levels (point fitting processing). This amplification coefficient was also introduced in the double impulse (Kojima and Takewaki 2016). It should be remarked that

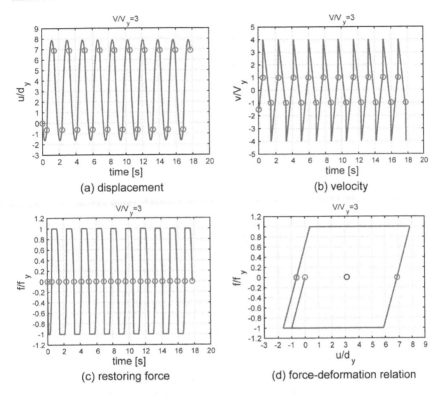

Figure 4.10 Response to Input Sequence 2 for V/V_y = 3: (a) displacement, (b) velocity, (c) restoring force, (d) force-deformation relation (Kojima and Takewaki 2015c).

the information on critical timing shown in Figure 4.5 was incorporated in Figure 4.12.

Figure 4.4 shows the comparison of the maximum plastic deformation u_p/d_y and the residual displacement u_r/d_y of the undamped EPP SDOF model under the multiple impulse and the corresponding multi-cycle sinusoidal wave with respect to the input velocity level. The multiple impulse provides a fairly good substitute of the multi-cycle sinusoidal wave in the evaluation of the maximum plastic deformation u_p/d_y if the amplitudes of both inputs are adjusted. Although the residual displacement exhibits a rather good correspondence between the multiple impulse (Input Sequence 1) and the corresponding multi-cycle sinusoidal wave, the Input Sequence 2 shows the somewhat larger residual displacement. However, since the Input Sequence 2 is used mainly for finding the critical timing, this discrepancy does not cause any problem.

Figure 4.12 presents the comparison of displacements to the multiple impulse (Input Sequence 1) and the corresponding sinusoidal wave for the

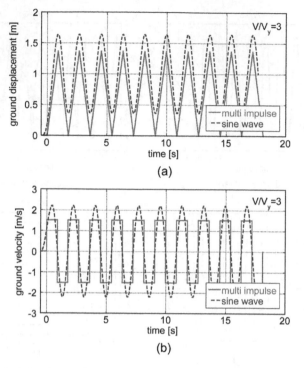

Figure 4.11 Comparison of ground displacement and velocity between multiple impulse and corresponding multi-cycle sinusoidal wave for input level $V/V_y = 3$ (Kojima and Takewaki 2015c).

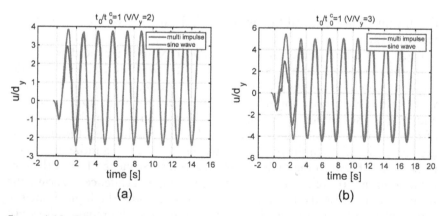

Figure 4.12 Comparison of responses to multiple impulse and sinusoidal wave (phase lag has been adjusted): (a) $V/V_y = 2$, (b) $V/V_y = 3$ (Kojima and Takewaki 2015c).

input velocity levels V/V_y = 2, 3 and the resonant impulse timing t_0/t_0^c = 1.0. As pointed out above, since the steady state is very sensitive to the time increment in the time-history response analysis using an EPP model due to the plastic flow, the time increment was chosen as 1.0×10^{-6}[sec]. This time increment requires a lot of computational time and load. On the other hand, the closed-form expression derived in this chapter provides an extremely efficient and reliable tool for response evaluation. In Figure 4.12, the structural parameters ω_1 = 2π[rad/sec] (T_1 = 1.0[sec]) and d_y = 0.16[m] were used. The phase lag was adjusted for the comparison purpose. The ground displacement and velocity of the corresponding sinusoidal wave for V/V_y = 3 are shown in Figure 4.11. Although a slight difference exists in the first cycle, both responses show a fairly good correspondence in the steady state. If desired, the residual displacement can be evaluated from Figure 4.4(b). As is well known, the residual displacement is sensitive to the irregularity in the input in the case of the EPP system. This issue is beyond the scope of this chapter.

4.6 PROOF OF CRITICAL TIMING

Figure 4.13 illustrates the normalized plastic deformation amplitude u_p/d_y with respect to the timing of multiple impulse input for various input velocity levels V/V_y=1, 2, 3, 4, 5 (Input Sequence 1). It can be understood that the critical timing t_0 = t_0^c derived from the Input Sequence 2 provides the critical timing even under Input Sequence 1. Repetitive appearance of peaks with

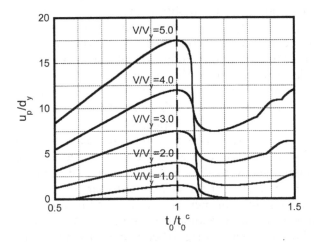

Figure 4.13 Plastic deformation amplitude with respect to timing of multiple impulse input for various input levels (Input Sequence 1).

the same amplitude indicates the existence of multiple solutions. However, the lowest (shortest) timing $t_0/t_0^c = 1.0$ may be meaningful from the viewpoint of occurrence possibility of such ground motion with long duration. The peak at the value of t_0 larger than t_0^c ($t_0/t_0^c > 1.0$) implies the action of the second impulse after the point with the zero restoring-force.

4.7 SUMMARIES

The obtained results may be summarized as follows:

1. The multiple impulse input of equal time interval can be a good substitute for the long-duration earthquake ground motion which is expressed in terms of harmonic waves. A closed-form expression can be derived for the maximum response of an elastic–perfectly plastic (EPP) single-degree-of-freedom (SDOF) model under the critical multiple impulse input.

2. While the critical set of input velocity amplitude and input frequency (timing of impulse) has to be computed iteratively for the multi-cycle sinusoidal wave, it can be obtained directly without iteration for the multiple impulse input by introducing a modified version (only the timing between the first and second impulses is modified so that the second impulse is given at the zero restoring force). The resonance was proved by using energy investigation. It was made clear that the critical timing of the multiple impulses can be characterized by the time attaining the zero restoring force in the unloading process. This decomposition of input amplitude and input frequency overcame the difficulty in finding the resonant frequency without repetition. This is one of the most original contributions in this chapter.

3. Since only free-vibration is induced under such multiple impulse input, a simple energy approach plays an important role in the derivation of the closed-form expression for a complicated elastic-plastic critical response. The energy approach enabled the derivation of the maximum critical elastic-plastic response without resorting to solving the equation of motion. The maximum elastic-plastic response after each impulse can be obtained by equating the initial kinetic energy given by the initial velocity to the sum of hysteretic dissipation and elastic strain energies. The critical inelastic deformation and the corresponding critical input frequency can be captured by the substituted multiple impulse input depending on the input velocity level. This is the second one of the most original contributions in this chapter.

4. The validity and accuracy of the theory explained in this chapter were assured through the comparison with the time-history response to the corresponding equivalent multi-cycle sinusoidal input. It was found that if the adjustment of both inputs is made by using the equivalence

of the maximum Fourier amplitude and a modification based on the response correspondence at some points with different input levels for better correspondence in a wider range of input level, the maximum elastic-plastic deformation to the multiple impulse exhibits a good correspondence with that to the multi-cycle sinusoidal wave.

5. While the conventional methods (Caughey 1960a, b, Iwan 1961, 1965a, b) are aimed at constructing an equivalent linear structural model or solving transcendental equations iteratively to an unchanged input (sinusoidal input) to enable the simple approximate computation of complicated elastic-plastic responses, the method explained in the present chapter is aimed at finding an equivalent input model for an unchanged exact elastic-plastic model. The most significant difference between two approaches is that while the conventional methods require the repetition both in the computation of equivalent model parameters or the solution of transcendental equations for one input frequency and the computation of the resonant frequency giving the maximum response, the approach explained in this chapter does not require any repetition in the computation of the critical input timing (resonant frequency) and the critical response. The approach explained in this chapter also enabled the computation of the steady-state response for an EPP model that cannot be treated by the conventional approaches.

The approach for an EPP model can be extended to a bilinear hysteretic model with some modifications (see Chapter 6).

REFERENCES

Abbas, A. M., and Manohar, C. S. (2002). Investigations into critical earthquake load models within deterministic and probabilistic frameworks, *Earthquake Eng. Struct. Dyn.*, 31, 813–832.

Akehashi, H., Kojima, K., Noroozinejad Farsangi, E., and Takewaki, I. (2018). Critical response evaluation of damped bilinear hysteretic SDOF model under long duration ground motion simulated by multi impulse motion, *Int. J. of Earthquake and Impact Eng*, 2(4), 298–321.

Akehashi, H., and Takewaki, I. (2020). Critical response of nonlinear base-isolated building considering soil-structure interaction, *Proc. of the 17th World Conference on Earthquake Engineering* (17WCEE), Sendai, Japan, September 13–18 (postponed).

Alavi, B., and Krawinkler, H. (2004). Behaviour of moment resisting frame structures subjected to near-fault ground motions, *Earthquake Eng. Struct. Dyn.*, 33(6), 687–706.

Bertero, V. V., Mahin, S. A., and Herrera, R. A. (1978). Aseismic design implications of near-fault San Fernando earthquake records, *Earthquake Eng. Struct. Dyn.* 6(1), 31–42.

Bozorgnia, Y., and Campbell, K. W. (2004). Engineering characterization of ground motion, Chapter 5 in *Earthquake engineering from engineering seismology to performance-based engineering*, (eds.) Bozorgnia, Y. and Bertero, V. V., CRC Press, Boca Raton, FL.

Bray, J. D., and Rodriguez-Marek, A. (2004). Characterization of forward-directivity ground motions in the near-fault region, *Soil Dyn. Earthquake Eng.*, 24(11), 815–828.

Caughey, T. K. (1960a). Sinusoidal excitation of a system with bilinear hysteresis, *J. Appl. Mech.*, 27(4), 640–643.

Caughey, T. K. (1960b). Random excitation of a system with bilinear hysteresis, *J. Appl. Mech.*, 27(4), 649–652.

Drenick, R. F. (1970). Model-free design of aseismic structures, *J. Eng. Mech. Div.*, ASCE, 96(EM4), 483–493.

Hall, J. F., Heaton, T. H., Halling, M. W., and Wald, D. J. (1995). Near-source ground motion and its effects on flexible buildings, *Earthquake Spectra*, 11(4), 569–605.

Hayashi, K., Fujita, K., Tsuji, M., and Takewaki, I. (2018). A simple response evaluation method for base-isolation building-connection hybrid structural system under long-period and long-duration ground motion, *Frontiers in Built Environment* (Specialty Section: Earthquake Engineering), Volume 4: Article 2.

Hayden, C. P., Bray, J. D., and Abrahamson, N. A. (2014). Selection of near-fault pulse motions, *J. Geotechnical Geoenvironmental Eng.*, ASCE, 140(7).

Iwan, W. D. (1961). *The dynamic response of bilinear hysteretic systems*, Ph.D. Thesis, California Institute of Technology.

Iwan, W. D. (1965a). The dynamic response of the one-degree-of-freedom bilinear hysteretic system, *Proceedings of the Third World Conference on Earthquake Engineering*, 1965.

Iwan, W. D. (1965b). The steady-state response of a two-degree-of-freedom bilinear hysteretic system, *Journal of Applied Mechanics*, 32(1), 151–156.

Kalkan, E., and Kunnath, S. K. (2006). Effects of fling step and forward directivity on seismic response of buildings, *Earthquake Spectra*, 22(2), 367–390.

Kalkan, E., and Kunnath, S. K. (2007). Effective cyclic energy as a measure of seismic demand, *J. Earthquake Eng.* 11, 725–751.

Kawai, A., Maeda, T., and Takewaki, I. (2020). Smart seismic control system for high-rise buildings using large-stroke viscous dampers through connection to strong-back core frame, *Frontiers in Built Environment* (Specialty Section: Earthquake Engineering), Volume 6, Article 29.

Khaloo, A. R., Khosravi, H., and Hamidi Jamnani, H. (2015). Nonlinear interstory drift contours for idealized forward directivity pulses using "Modified Fish-Bone" models; *Advances in Structural Eng.*18(5), 603–627.

Kojima, K., Fujita, K., and Takewaki, I. (2015). Critical double impulse input and bound of earthquake input energy to building structure, *Frontiers in Built Environment*, Volume 1, Article 5.

Kojima, K., and Takewaki, I. (2015a). Critical earthquake response of elastic-plastic structures under near-fault ground motions (Part 1: Fling-step input), *Frontiers in Built Environment*, Volume 1, Article 12.

Kojima, K., and Takewaki, I. (2015b). Critical earthquake response of elastic-plastic structures under near-fault ground motions (Part 2: Forward-directivity input), *Frontiers in Built Environment*, Volume 1, Article 13.

Kojima, K., and Takewaki, I. (2015c). Critical input and response of elastic-plastic structures under long-duration earthquake ground motions, *Frontiers in Built Environment* (Specialty Section: Earthquake Engineering), Volume 1, Article 15.

Kojima, K., and Takewaki, I. (2016). Closed-form critical earthquake response of elastic-plastic structures on compliant ground under near-fault ground motions, *Frontiers in Built Environment* (Specialty Section: Earthquake Engineering), Volume 2, Article 1.

Kojima, K., and Takewaki, I. (2017). Critical steady-state response of SDOF bilinear hysteretic system under multi impulse as substitute of long-duration ground motions, *Frontiers in Built Environment* (Specialty Section: Earthquake Engineering), Volume 3: Article 41.

Liu, C.-S. (2000). The steady loops of SDOF perfectly elastoplastic structures under sinusoidal loadings, *J. Marine Science and Technology*, 8(1), 50–60.

Luco, J. E. (2014). Effects of soil-structure interaction on seismic base isolation, *Soil Dyn. Earthq. Eng.*, 66, 67–177. doi: 10.1016/j.soildyn.2014.05.007

Mavroeidis, G. P., and Papageorgiou, A. S. (2003). A mathematical representation of near-fault ground motions, *Bull. Seism. Soc. Am.*, 93(3), 1099–1131.

Mavroeidis, G. P., Dong, G., and Papageorgiou, A. S. (2004). Near-fault ground motions, and the response of elastic and inelastic single-degree-freedom (SDOF) systems, *Earthquake Eng. Struct. Dyn.*, 33, 1023–1049.

Moustafa, A., Ueno, K., and Takewaki, I. (2010). Critical earthquake loads for SDOF inelastic structures considering evolution of seismic waves, *Earthquakes and Structures*, 1(2), 147–162.

Mukhopadhyay, S., and Gupta, V. K. (2013a). Directivity pulses in near-fault ground motions—I: Identification, extraction and modeling, *Soil Dyn. Earthquake Eng.*, 50, 1–15.

Mukhopadhyay, S., and Gupta, V. K. (2013b). Directivity pulses in near-fault ground motions—II: Estimation of pulse parameters, *Soil Dyn. Earthquake Eng.*, 50, 38–52.

Roberts, J. B., and Spanos, P. D. (1990). *Random vibration and statistical linearization*. Wiley, New York.

Rupakhety, R., and Sigbjörnsson, R. (2011). Can simple pulses adequately represent near-fault ground motions?, *J. Earthquake Eng.*, 15,1260–1272.

Sasani, M., and Bertero, V. V. (2000). Importance of severe pulse-type ground motions in performance-based engineering: historical and critical review, in *Proceedings of the Twelfth World Conference on Earthquake Engineering*, Auckland, New Zealand.

Takewaki, I. (1996). Design-oriented approximate bound of inelastic responses of a structure under seismic loading, *Computers and Structures*, 61(3), 431–440.

Takewaki, I. (1997). Design-oriented ductility bound of a plane frame under seismic loading, *J. Vibration and Control*, 3(4), 411–434.

Takewaki, I. (2002). Robust building stiffness design for variable critical excitations, *J. Struct. Eng.*, ASCE, 128(12), 1565–1574.

Takewaki, I. (2004). Bound of earthquake input energy, *J. Struct. Eng.*, ASCE, 130(9), 1289–1297.

Takewaki, I. (2007). *Critical excitation methods in earthquake engineering*, Oxford, Elsevier (Second edition in 2013).

Takewaki, I., and Fujita, K. (2009). Earthquake input energy to tall and base-isolated buildings in time and frequency dual domains, *J. of The Structural Design of Tall and Special Buildings*, 18(6), 589–606.

Takewaki, I., Murakami, S., Fujita, K., Yoshitomi, S., and Tsuji, M. (2011). The 2011 off the Pacific coast of Tohoku earthquake and response of high-rise buildings under long-period ground motions, *Soil Dyn. Earthquake Eng.*, 31(11), 1511–1528.

Takewaki, I., and Tsujimoto, H. (2011). Scaling of design earthquake ground motions for tall buildings based on drift and input energy demands, *Earthquakes Struct.*, 2(2), 171–187.

Takewaki, I., Moustafa, A., and Fujita, K. (2012). *Improving the earthquake resilience of buildings: The worst case approach*, Springer, London.

Tamura, G., Kojima, K., and Takewaki, I. (2019). Critical response of elastic-plastic SDOF systems with nonlinear viscous damping under simulated earthquake ground motions, *Heliyon*, 5, e01221. doi:10.1016/j.heliyon.2019.e01221.

Thomson, W. T. (1965). *Vibration theory and applications*, Prentice-Hall, Englewood Cliffs, NJ.

Vafaei, D., and Eskandari, R. (2015). Seismic response of mega buckling-restrained braces subjected to fling-step and forward-directivity near-fault ground motions, *Struct. Design Tall Spec. Build.*, 24, 672–686.

Xu, Z., Agrawal, A. K., He, W.-L., and Tan, P. (2007). Performance of passive energy dissipation systems during near-field ground motion type pulses, *Eng. Struct.*, 29, 224–236.

Yamamoto, K., Fujita, K., and Takewaki, I. (2011). Instantaneous earthquake input energy and sensitivity in base-isolated building, *Struct. Design Tall Spec. Build.*, 20(6), 631–648.

Yang, D., and Zhou, J. (2014). A stochastic model and synthesis for near-fault impulsive ground motions, *Earthquake Eng. Struct. Dyn.*, 44, 243–264.

Zhai, C., Chang, Z., Li, S., Chen, Z.-Q., and Xie, L. (2013). Quantitative identification of near-fault pulse-like ground motions based on energy, *Bull. Seism. Soc. Am.*, 103(5), 2591–2603.

A. APPENDIX 1: MULTI-IMPULSE AND CORRESPONDING MULTI-CYCLE SINE WAVE WITH THE SAME FREQUENCY AND SAME MAXIMUM FOURIER AMPLITUDE

The velocity amplitude V of the multi-impulse is related to the maximum velocity of the corresponding multi-cycle sine wave with the same frequency (the period is twice the interval of the multi impulse) so that the maximum Fourier amplitudes of both inputs coincide. The detail is explained in this section.

The multi-impulse is expressed by

$$\ddot{u}_g(t) = 0.5V\delta(t) - V\delta(t - t_0) + V\delta(t - 2t_0) - V\delta(t - 3t_0) + \cdots$$
$$+ (-1)^{N-1} V\delta\{t - (N-1)t_0\} + 0.5V\delta(t - Nt_0) \tag{4.A1}$$

where V is the velocity amplitude of the multi-impulse and $\delta(t)$ is the Dirac delta function. The Fourier transform of Eq. (4.A1) can be obtained as

$$\ddot{U}_g(\omega) = \int_{-\infty}^{\infty} \left\{ \begin{array}{l} 0.5V\delta(t) - V\delta(t-t_0) + V\delta(t-2t_0) - V\delta(t-3t_0) + \cdots \\ +(-1)^{N-1} V\delta\{t-(N-1)t_0\} + 0.5V\delta(t-Nt_0) \end{array} \right\} e^{-i\omega t} dt$$

$$= \int_{-\infty}^{\infty} \left\{ \begin{array}{l} 0.5V\delta(t)e^{-i\omega t} - V\delta(t-t_0)e^{-i\omega t_0}e^{-i\omega(t-t_0)} \\ +V\delta(t-2t_0)e^{-i\omega 2t_0}e^{-i\omega(t-2t_0)} - V\delta(t-3t_0)e^{-i\omega 3t_0}e^{-i\omega(t-3t_0)} + \cdots \\ +(-1)^{N-1}V\delta\{t-(N-1)t_0\}e^{-i\omega(N-1)t_0}e^{-i\omega\{t-(N-1)t_0\}} \\ +0.5V\delta(t-Nt_0)e^{-i\omega Nt_0}e^{-i\omega(t-Nt_0)} \end{array} \right\} dt \quad (4.A2)$$

$$= V\left\{ 0.5 - e^{-i\omega t_0} + e^{-i\omega 2t_0} - e^{-i\omega 3t_0} + \cdots + (-1)^{N-1}e^{-i\omega(N-1)t_0} + 0.5e^{-i\omega Nt_0} \right\}$$

Let A_l, T_l, and $\omega_l = 2\pi/T_l$ denote the acceleration amplitude, the period, and the circular frequency of the corresponding multi-cycle sine wave, respectively. The acceleration wave $\ddot{u}_g^{SW}$ of the corresponding multi-cycle sine wave is expressed by

$$\ddot{u}_g^{SW}(t) = \begin{cases} 0.5A_l\sin(\omega_l t) & (0 \le t < 0.5T_l, 0.5NT_l < t \le 0.5(N+1)T_l) \\ A_l\sin(\omega_l t) & (0.5T_l \le t \le 0.5NT_l) \end{cases} \quad (4.A3)$$

The time interval t_0 of consecutive two impulses in the multi-impulse is related to the period T_l of the corresponding multi-cycle sine wave by $T_l = 2t_0$. Although the starting points of both inputs differ by $t_0/2$, the starting time of multi-cycle sine wave does not affect the Fourier amplitude. For this reason, the starting time of multi-cycle sine wave will be adjusted so that the responses of both inputs correspond well. In this section, the relation of the velocity amplitude V of the multi-impulse with the acceleration amplitude A_l of the corresponding multi-cycle sine wave is derived. The ratio $a_M^b(t_0, N)$ of A_l to V is introduced by

$$A_l = a_M^b(t_0, N)V \quad (4.A4)$$

The Fourier transform of $\ddot{u}_g^{SW}$ in Eq.(4.A3) is computed by

$$\ddot{U}_g^{SW}(\omega) = 0.5\int_0^{0.5T_l} \{A_l\sin(\omega_l t)\} e^{-i\omega t} dt + \int_{0.5T_l}^{0.5NT_l} \{A_l\sin(\omega_l t)\} e^{-i\omega t} dt$$

$$+0.5\int_{0.5NT_l}^{0.5(N+1)T_l} \{A_l\sin(\omega_l t)\} e^{-i\omega t} dt \quad (4.A5)$$

$$= -A_l\pi t_0 \left\{ -0.5 + 0.5e^{-i\omega t_0} + 0.5e^{-i\omega Nt_0} - 0.5e^{-i\omega(N+1)t_0} \right\} / \left\{ \pi^2 - (\omega t_0)^2 \right\}$$

From Eqs. (4.A2), (4.A5), the Fourier amplitudes of both inputs are expressed by

$$\left|\ddot{U}_g(\omega)\right| = V\left|0.5 + \sum_{n=1}^{N-1}(-1)^n e^{-i\omega n t_0} + 0.5 e^{-i\omega N t_0}\right| \quad (4.A6)$$

$$\left|\ddot{U}_g^{SW}(\omega)\right| = A_l \pi t_0 \left|\left\{-0.5 + 0.5 e^{-i\omega t_0} + 0.5 e^{-i\omega N t_0} - 0.5 e^{-i\omega(N+1)t_0}\right\} / \left\{\pi^2 - (\omega t_0)^2\right\}\right| \quad (4.A7)$$

The coefficient $a_M{}^b(t_0, N)$ can be derived from Eqs. (4.A4), (4.A6), (4.A7) and the equivalence of the maximum Fourier amplitude $\left|\ddot{U}_g(\omega)\right|_{max} = \left|\ddot{U}_g^{SW}(\omega)\right|_{max}$.

$$a_M{}^b(t_0, N) = \frac{A_l}{V} = \frac{\max\left|0.5 + \sum_{n=1}^{N-1}(-1)^n e^{-i\omega n t_0} + 0.5 e^{-i\omega N t_0}\right|}{\max\left|\pi t_0\left\{0.5 - 0.5 e^{-i\omega t_0} - 0.5 e^{-i\omega N t_0} + 0.5 e^{-i\omega(N+1)t_0}\right\} / \left\{\pi^2 - (\omega t_0)^2\right\}\right|} \quad (4.A8)$$

In the numerator of Eq. (4.A8), $\max\left|0.5 + \sum_{n=1}^{N-1}(-1)^n e^{-i\omega n t_0} + 0.5 e^{-i\omega N t_0}\right| = N$ holds. The denominator of Eq. (4.A8) will be evaluated next.

Let us define the function $f_M{}^b(x, N)$ given by

$$f_M{}^b(x, N) = \left|\frac{0.5 - 0.5 e^{-ix} - e^{-iNx} + 0.5 e^{-i(N+1)x}}{\pi^2 - x^2}\right| \quad (4.A9)$$

When we substitute $x = \omega t_0$ in Eq. (4.A9), Eq. (4.A7) can be transformed into

$$\left|\ddot{U}_g^{SW}(\omega)\right| / A_l = (\pi t_0) f_M{}^b(x = \omega t_0, N) \quad (4.A10)$$

Eq. (4.A9) means that once the maximum value of $f_M{}^b(x = \omega t_0, N)$ is obtained, the denominator of Eq. (4.A8) is evaluated. After some manipulation on the maximization of Eq. (4.A10), the following relation is derived.

$$V_l = A_l / \omega_l = NV / \left\{\pi^2 f_{M\max}{}^b(N)\right\} \quad (4.A11)$$

where $f_{M\max}{}^b(N)$ is the maximum value of $f_M{}^b(x, N)$ for the variable x. Finally, for a sufficiently large number $N(\geq 20)$, the function $f_M{}^b(x, N)$ becomes maximum at $x = \pi$ and $f_M{}^b(x = \pi, N) = N/(2\pi)$. Then, Eq. (4.A11) is reduced to

$$V_l = A_l/\omega_l = (2/\pi)V \qquad\qquad (4.A12)$$

Eq. (4.A12) provides the relation between V and A_l in terms of $t_0(=T_l/2)$ together with the relation between V and V_l.

Chapter 5

Critical earthquake response of an elastic–perfectly plastic SDOF model with viscous damping under double impulse

5.1 INTRODUCTION

In Chapter 1, the earthquake energy balance law was introduced for a damped single-degree-of-freedom (SDOF) model (Eq. (1.28) in Section 1.3.2). It was pointed out there that the energy dissipated by the viscous damping has to be evaluated in terms of the maximum displacement in an appropriate manner even if approximate in order to obtain the maximum displacement from the energy balance law. In this chapter, the critical earthquake response of an elastic–perfectly plastic (EPP) SDOF model with viscous damping under the double impulse is derived in an explicit manner. The quadratic-function approximation for the damping force-deformation relationship is introduced and enables the application of the earthquake energy balance law to the damped model. The validity and accuracy of the proposed theory are investigated through comparison of the results of the nonlinear time-history response analysis to the corresponding one-cycle sinusoidal input and actual recorded ground motions.

5.2 MODELING OF NEAR-FAULT GROUND MOTION WITH DOUBLE IMPULSE

In this chapter, the double impulse is used as in Chapters 1 and 2. The critical time interval in the double impulse for the undamped elastic–perfectly plastic (EPP) SDOF system, which corresponds to a half of the resonant period in the one-cycle sinusoidal wave, can be obtained directly. In contrast, the resonant frequency of the one-cycle sinusoidal wave has to be computed for a specified input level by changing the input frequency parametrically and transforming the structural model to an equivalent model or solving complicated transcendental equations in the conventional approach (Caughey 1960, Iwan 1961).

109

A ground acceleration $\ddot{u}_g(t)$ in terms of the double impulse is expressed by

$$\ddot{u}_g(t) = V\delta(t) - V\delta(t - t_0) \tag{5.1}$$

where V is the velocity amplitude of the double impulse, $\delta(t)$ is the Dirac delta function, and t_0 is the time interval of two impulses. As shown in Chapter 1, the acceleration $\ddot{u}_g^{SW}$ of the corresponding one-cycle sine wave is expressed by

$$\ddot{u}_g^{SW} = A_p \sin(\omega_p t) \quad (0 \le t \le T_p = 2t_0) \tag{5.2}$$

where A_p is the acceleration amplitude and ω_p is the circular frequency. The Fourier transform of Eq. (5.1) leads to

$$\ddot{U}_g(\omega) = V\left(1 - e^{-i\omega t_0}\right) \tag{5.3}$$

On the other hand, the Fourier transform of Eq. (5.2) was presented in Section 1.2 (Eq. (1.5)).

5.3 ELASTIC–PERFECTLY PLASTIC SDOF MODEL WITH VISCOUS DAMPING

Consider a damped EPP SDOF model of mass m, stiffness k, and damping coefficient c as shown in Figure 5.1. It is assumed that the damping coefficient does not change, regardless of yielding. The parameters $\omega_1 = \sqrt{k/m}$, $T_1 = 2\pi/\omega_1$, and $h = c/\left(2\sqrt{km}\right)$ denote the undamped natural circular frequency, the undamped natural period, and the damping ratio, respectively. On the other hand, the parameters $\omega_1' = \sqrt{1 - h^2}\,\omega_1$ and $T_1' = 2\pi/\omega_1'$ denote the damped natural circular frequency and the damped natural period, respectively. As for deformation and force parameters, let u, f_R and f_D denote the

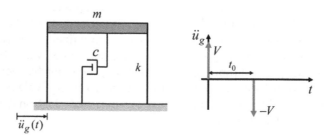

Figure 5.1 Elastic-perfectly plastic SDOF model with viscous damping under double impulse.

displacement of the mass relative to the ground (deformation of the system), the restoring force in the spring, and the damping force in the dashpot, respectively. The parameters d_y and f_y denote the yield deformation and the yield force, respectively. The deformation and force parameters will be treated as normalized parameters to capture the intrinsic relation between the input parameters and the elastic-plastic response.

5.4 ELASTIC-PLASTIC RESPONSE OF UNDAMPED SYSTEM TO CRITICAL DOUBLE IMPULSE

Kojima and Takewaki (2015a) derived a closed-form expression on the maximum elastic-plastic responses of an undamped EPP SDOF model under the critical double impulse (see Chapter 2). Those responses can be derived by an energy balance approach without solving the equation of motion directly. More specifically, the maximum deformation can be calculated by using the energy balance law, in which the kinetic energies given at the times of the first impulse and the second impulse are transformed into the sum of the elastic strain energy corresponding to the yield deformation and the energy dissipated during the plastic deformation (see Figures 1.5 and 1.7). The critical elastic-plastic response can be derived in closed form, and the critical time interval (corresponding to a half of the resonant period) can be derived automatically for the increasing input velocity level of the double impulse by using this method. Since it is expected that a similar theory can be developed for deriving the maximum elastic-plastic response of a damped SDOF system, the closed-form expression for the maximum deformation of the undamped EPP SDOF model that was derived in Kojima and Takewaki (2015a) and Chapter 2 is explained briefly in this section for later smooth connection.

The maximum deformations after the first and second impulses are denoted by u_{max1} and u_{max2}, respectively, as shown in Figure 5.2, and the

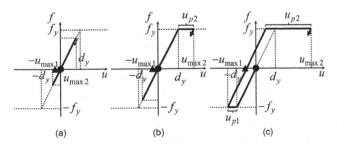

Figure 5.2 Maximum deformation of elastic-perfectly plastic model to critical double impulse: (a) CASE 1: elastic range, (b) CASE 2: yielding after 2nd impulse, (c) CASE 3: yielding after 1st impulse (●: 1st impulse, ▲: 2nd impulse) (Kojima et al. 2017, 2018).

maximum deformation under the critical double impulse is evaluated by $u_{max} = \max(u_{max1}, u_{max2})$. Note that both parameters u_{max1} and u_{max2} are the absolute values. The plastic deformations after the first and second impulses are denoted by u_{p1} and u_{p2}, respectively. The maximum elastic-plastic response of the undamped EPP SDOF model under the critical double impulse can be classified into one of three cases, depending on the input velocity level (yielding stage). CASE 1 is the case of the elastic response even after the second impulse. CASE 2 is the case of the plastic deformation only after the second impulse. Finally, CASE 3 is the case of the plastic deformation after the first impulse. Figure 5.2 shows a schematic diagram of CASE 1, CASE 2, and CASE 3.

Let $V_y(=\omega_1 d_y)$ denote the input velocity level of the double impulse at which the maximum deformation of the undamped EPP SDOF model just attains the yield deformation after the first impulse. It is important to note that this parameter can be regarded as a strength parameter of the undamped EPP SDOF model. This parameter V_y is used for normalizing the input velocity level, and V/V_y is simply called the input velocity level. The maximum deformations u_{max1} and u_{max2} with respect to V/V_y in CASES 1–3 can be obtained as follows by using the energy balance law.

$$
\frac{u_{max1}}{d_y} = \begin{cases} V/V_y & \text{for} \quad 0 \le V/V_y < 1.0 \quad (\text{CASEs 1, 2}) \\ 0.5\left\{1 + \left(V/V_y\right)^2\right\} & \text{for} \quad 1.0 \le V/V_y \qquad (\text{CASE 3}) \end{cases} \quad (5.4)
$$

$$
\frac{u_{max2}}{d_y} = \begin{cases} 2V/V_y & \text{for} \quad 0 \le V/V_y < 0.5 \quad (\text{CASE 1}) \\ 0.5\left\{1 + \left(2V/V_y\right)^2\right\} & \text{for} \quad 0.5 \le \dfrac{V}{V_y} < 1.0 \quad (\text{CASE 2}) \\ 1.5 + \left(V/V_y\right) & \text{for} \quad 1.0 \le V/V_y \qquad (\text{CASE 3}) \end{cases} \quad (5.5)
$$

Figure 5.3 shows the maximum deformation of the undamped EPP SDOF model under the critical double impulse normalized by the yield deformation with respect to input velocity level V/V_y. The critical timing of the second impulse (the critical time interval), which maximizes the maximum deformation u_{max2} after the second impulse under a constant level of V/V_y, is characterized as the time when the restoring force attains zero in the unloading process after the first impulse (Kojima and Takewaki 2015a and Chapter 2). In CASES 1 and 2, since the response after the first impulse is in an elastic range, the critical time interval $t_0{}^c$ is a half of the initial natural period T_1 of the undamped EPP SDOF model. In CASE 3, since the undamped EPP SDOF model enters the yielding stage after the first impulse, it is necessary to derive the expression by solving the equation of motion. The critical time interval $t_0{}^c$ can be obtained by solving the equation of motion as follows.

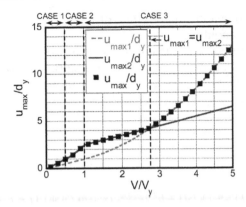

Figure 5.3 Maximum deformation u_{max}/d_y for input level V/V_y (Kojima and Takewaki 2015a).

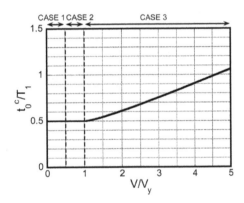

Figure 5.4 Critical impulse timing $t_0{}^c/T_1$ for input level V/V_y (Kojima and Takewaki 2015a).

$$\frac{t_0^c}{T_1} = \begin{cases} 0.5 & \text{for} \quad 0 \le V/V_y < 1.0 \quad \text{(CASEs 1, 2)} \\ \left\{ \arcsin\left(V_y/V\right) + \sqrt{\left(V/V_y\right)^2 - 1} \right\} / \left(2\pi\right) + 1/4 & \text{for } 1.0 \le V/V_y \quad \text{(CASE 3)} \end{cases} \tag{5.6}$$

Figure 5.4 illustrates the critical time interval t_0^c, normalized by T_1, with respect to input velocity level V/V_y.

5.5 LINEAR ELASTIC RESPONSE OF DAMPED SYSTEM TO CRITICAL DOUBLE IMPULSE

In this section, a closed-form expression is derived for the maximum deformation of an elastic SDOF system with viscous damping under the critical double impulse to investigate the effect of viscous damping on the response under the double impulse.

Note that the response of the linear elastic SDOF system after the second impulse can be obtained by simply superposing the free vibrations after the first and second impulses. Since the velocity input of the first and second inputs are expressed by V and $-V$, the deformation and velocity responses after the first and second impulses can be obtained with respect to general time interval t_0 as follows.

$$u(t) = -\left(V/\omega_1'\right)e^{-h\omega_1 t}\sin\left(\omega_1' t\right) \ \left(t < t_0\right), \tag{5.7a}$$

$$\dot{u}(t) = -\left(V/\sqrt{1-h^2}\right)e^{-h\omega_1 t}\cos\left(\omega_1' t + \phi\right) \ \left(t < t_0\right), \tag{5.7b}$$

$$
\begin{aligned}
u(t) &= -\frac{V}{\omega_1'}e^{-h\omega_1 t}\sin\left(\omega_1' t\right) + \frac{V}{\omega_1'}e^{-h\omega_1(t-t_0)}\sin\left\{\omega_1'\left(t-t_0\right)\right\} \\
&= -\frac{V}{\omega_1'}\sqrt{1-2e^{h\omega_1 t_0}\cos\left(\omega_1' t_0\right)+e^{2h\omega_1 t_0}}\,e^{-h\omega_1 t}\sin\left(\omega_1' t + \theta\right) \ \left(t \geq t_0\right),
\end{aligned}
\tag{5.8a}
$$

$$\dot{u}(t) = -V\sqrt{\frac{1-2e^{h\omega_1 t_0}\cos\left(\omega_1' t_0\right)+e^{2h\omega_1 t_0}}{1-h^2}}\,e^{-h\omega_1 t}\cos\left(\omega_1' t + \theta + \phi\right) \ \left(t \geq t_0\right), \tag{5.8b}$$

where

$$\phi = \arctan\left(h/\sqrt{1-h^2}\right) \tag{5.9a}$$

$$
\theta = \begin{cases}
\arctan\dfrac{e^{h\omega_1 t_0}\sin\left(\omega_1' t_0\right)}{1-\exp\left(h\omega_1 t_0\right)\cos\left(\omega_1' t_0\right)} & \left(1-e^{h\omega_1 t_0}\cos\left(\omega_1' t_0\right)\geq 0\right) \\[4mm]
\arccos\dfrac{1-e^{h\omega_1 t_0}\cos\left(\omega_1' t_0\right)}{\sqrt{1-2e^{h\omega_1 t_0}\cos\left(\omega_1' t_0\right)+e^{2h\omega_1 t_0}}} & \left(\begin{array}{l}1-e^{h\omega_1 t_0}\cos\left(\omega_1' t_0\right)<0 \\ \text{and} \quad e^{h\omega_1 t_0}\sin\left(\omega_1' t_0\right)\geq 0\end{array}\right) \\[4mm]
-\arccos\dfrac{1-e^{h\omega_1 t_0}\cos\left(\omega_1' t_0\right)}{\sqrt{1-2e^{h\omega_1 t_0}\cos\left(\omega_1' t_0\right)+e^{2h\omega_1 t_0}}} & \left(\begin{array}{l}1-e^{h\omega_1 t_0}\cos\left(\omega_1' t_0\right)<0 \\ \text{and} \quad e^{h\omega_1 t_0}\sin\left(\omega_1' t_0\right)<0\end{array}\right)
\end{cases}
\tag{5.9b}
$$

The deformation responses after the first and second impulses are maximized at the time satisfying $\dot{u} = 0$. The maximum deformations u_{max1} and u_{max2} (absolute values) after the first and second impulses, respectively, can be obtained as follows.

$$
u_{\text{max1}} = \begin{cases}
\left(V/\omega_1'\right)\exp\left(-h\omega_1 t_0\right)\sin\left(\omega_1' t_0\right) & \left(t_0 < t_{\text{max1}}\right) \\
\left(V/\omega_1'\right)\exp\left(-h\omega_1 t_{\text{max1}}\right)\sin\left(\omega_1' t_{\text{max1}}\right) & \left(t_0 \geq t_{\text{max1}}\right)
\end{cases}
\tag{5.10a}
$$

$$u_{max2} = \frac{-V}{\omega_1'} e^{-h\omega_1 t_{max2}} \sqrt{1 - 2e^{h\omega_1 t_0} \cos(\omega_1' t_0) + e^{2h\omega_1 t_0}} \sin(\omega_1' t_{max2} + \theta), \quad (5.10b)$$

where

$$t_{max1} = \{0.25 - \phi/(2\pi)\} T_1', \quad (5.11a)$$

$$t_{max2} = \{(4N-1)/4 - (\theta + \phi)/(2\pi)\} T_1'. \quad (5.11b)$$

In Eq. (5.11b), N is a positive integer satisfying $N - 1 \le (t_0/T_1') < N$.

From Eqs. (5.10a, b) and (5.11a, b), the relation between the time interval t_0 and the maximum deformation $u_{max} = \max(u_{max1}, u_{max2})$ can be obtained in an explicit manner. Figure 5.5 shows the maximum deformation $u_{max} = \max(u_{max1}, u_{max2})$ with respect to t_0 for various damping ratios $h = 0.01, 0.02, 0.05, 0.1, 0.2, 0.5$. The abscissa is t_0, which is normalized by the critical time interval t_0^c, and the ordinate is u_{max}, which is normalized by $2V/\omega_1$. This value $2V/\omega_1$ indicates the maximum deformation of the undamped elastic SDOF system under the critical double impulse. The critical time interval of the elastic SDOF system is $t_0^c = T_1'/2$. This can be proven by setting $du_{max2}/dt_0|_{t_0 = t_0^c} = 0$ (see Appendix 1), and the restoring force attains zero at $t = T_1'/2$ after the first impulse (the first impulse acts at $t = 0$).

It should be ensured that the velocity response of the damped SDOF system does not attain the maximum after the first impulse at the critical timing expressed by the zero restoring-force timing. The maximum deformation u_{max1}, u_{max2} of the elastic SDOF system with viscous damping under the critical double impulse can be obtained as follows by substituting $t_0 = T_1'/2$ into Eqs. (5.10a, b) and (5.11a, b).

$$u_{max1} = \frac{V}{\omega_1} \exp\left\{-\frac{h}{\sqrt{1-h^2}}\left(\frac{\pi}{2} - \arctan\frac{h}{\sqrt{1-h^2}}\right)\right\} \quad (5.12a)$$

$$u_{max2} = \frac{V}{\omega_1} \exp\left\{-\frac{h}{\sqrt{1-h^2}}\left(\frac{3}{2}\pi - \arctan\frac{h}{\sqrt{1-h^2}}\right)\right\}\left\{1 + \exp\left(\frac{\pi h}{\sqrt{1-h^2}}\right)\right\} \quad (5.12b)$$

Figure 5.6 shows a comparison of the maximum deformation under the double impulse with respect to the time interval t_0 and the maximum deformation under the corresponding one-cycle sinusoidal wave with respect to the input period T_p. It should be noted that $t_0 = T_p/2$ is used in the plot for the one-cycle sinusoidal wave. The damping ratios are taken as $h = 0.05, 0.2$ in Figure 5.6. The input period T_p is changed for the specific maximum velocity calculated by $V_p = 1.2222V$ with the input velocity level V. The critical input period T_p^c is double the critical time interval t_0^c because of the correspondence between the double impulse and the one-cycle sinusoidal

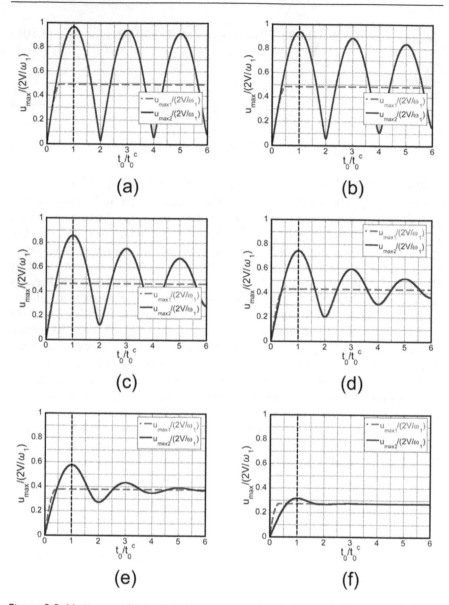

Figure 5.5 Maximum elastic deformation u_{max}/d_y with viscous damping to double impulse for varied impulse timing: (a) $h = 0.01$, (b) $h = 0.02$, (c) $h = 0.05$, (d) $h = 0.1$, (e) $h = 0.2$, (f) $h = 0.5$ (Kojima et al. 2017, 2018).

wave. As stated above, the abscissa $t_0/t_0^c(=T_p/T_p^c)$ for the corresponding sinusoidal wave denotes the input period T_p, normalized by the approximate critical period $T_p^c(=2t_0^c)$. Although the maximum deformation under the corresponding one-cycle sinusoidal wave is maximized at a period that is slightly shorter than the critical input period T_p^c that is calculated by using

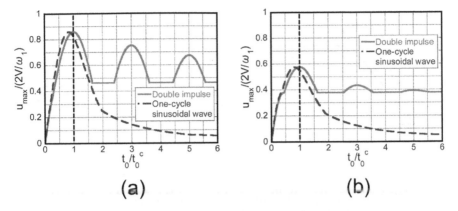

Figure 5.6 Comparison of maximum elastic deformation u_{max}/d_y with viscous damping to double impulse and equivalent one-cycle sine wave for varied impulse timing: (a) $h = 0.05$, (b) $h = 0.2$ (Kojima et al. 2017, 2018).

the critical double impulse, the maximum deformation under the critical double impulse is in good agreement with the upper bound of the maximum deformation under the corresponding one-cycle sinusoidal wave.

5.6 ELASTIC-PLASTIC RESPONSE OF DAMPED SYSTEM TO CRITICAL DOUBLE IMPULSE

5.6.1 Approximate critical response of the elastic-plastic system with viscous damping based on the energy balance law

In this section, a closed-form expression is derived for the maximum deformation of the EPP SDOF system with viscous damping under the critical double impulse. The maximum deformation of the undamped EPP SDOF system under the critical double impulse can be evaluated by using the energy balance law, in which the kinetic energies given at the time of the first impulse and the second impulse are transformed into the sum of the hysteretic energy and the maximum elastic strain energy corresponding to the yield deformation (see Section 1.3 in Chapter 1). On the other hand, in the EPP SDOF system with viscous damping, the kinetic energies given at the first impulse and the second impulse are equal to the sum of the elastic strain energy corresponding to the yield deformation, the energy dissipated during the plastic deformation and the work done by the damping force (the energy consumed by viscous damping). This corresponds to the energy balance law for the elastic-plastic system with viscous damping (see Figure 5.7). The essential feature was explained in Section 1.3.2 in Chapter 1.

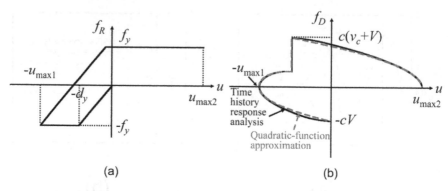

Figure 5.7 Quadratic approximation of damping force-deformation relation and its application to evaluation of maximum deformation of elastic–perfectly plastic model with viscous damping: (a) restoring force-deformation relation, (b) damping force-deformation relation (Kojima et al. 2017, 2018).

However, it is difficult to obtain the exact analytical expression for the response of the system with both hysteretic damping and viscous damping by solving the differential equation, even by using a simple input such as a double impulse. Although it may be possible to solve the equations of motion for both time intervals in the elastic range and the plastic range and determine the undetermined coefficients by using the initial and continuation conditions, its expression is too complicated. In this section, a method is developed and explained for approximating the damping force-deformation relation in terms of the maximum deformation by using a quadratic function whose vertex is at the zero-velocity point (the point of maximum deformation). A quadratic function that passes through both the acting point of the first or second impulse, and the zero-velocity point may be the simplest function that can represent the behavior of the damping force-deformation relation near the zero-velocity point. Using this approximation, the work done by the damping force can be represented by using the damping force that is calculated based on the initial velocity (the velocity just after the first or second impulse) and the maximum deformation.

According to Sections 5.4 and 5.5, the critical timing of the second impulse of both the linear elastic system with viscous damping and the undamped elastic-plastic system is the zero-restoring-force timing in the unloading process after the first impulse. Based on these observations, it may be appropriate to assume that the critical timing of the second impulse of the elastic-plastic SDOF system with viscous damping is also the zero-restoring-force timing in the unloading process after the first impulse. The validity of this assumption will be checked numerically in Section 5.6.5. In this sense, the approximate closed-form expression is derived in this section for the maximum deformation of the EPP SDOF system with viscous damping under the critical double impulse. The present formulation uses (1) the quadratic function approximation for the damping force-deformation relation, (2) the assumption that the zero-restoring-force timing is the critical timing of the second impulse, and (3) the energy balance law for the elastic-plastic system with viscous damping. Furthermore, the accuracy of the approximation (1) stated above and the assumption (2) stated above are investigated by using nonlinear time-history response analysis.

The maximum deformations after the first and second impulses are denoted by u_{max1} and u_{max2}, respectively, as shown in Figure 5.7, and the maximum deformation under the critical double impulse is evaluated by $u_{max} = \max(u_{max1}, u_{max2})$. Note that u_{max1} and u_{max2} are the absolute values. The elastic-plastic response of the EPP SDOF system with viscous damping under the critical double impulse can be classified into one of three cases, depending on the input velocity level. CASE 1 is the case of elastic response, even after the second impulse. CASE 2 is the case of plastic deformation only after the second impulse. Finally, CASE 3 is the case of plastic deformation after the first impulse. Figure 5.8 shows a schematic diagram of the restoring force characteristic and the damping force-deformation relation for CASE 1, CASE 2, and CASE 3. In this section, the parameter V_y is equal to $V_y(=\omega_1 d_y)$, as defined in Section 5.4, although the model does not attain the yield deformation exactly after the first impulse.

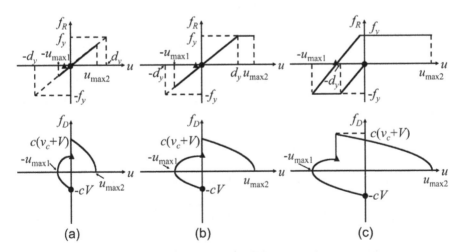

Figure 5.8 Evaluation of maximum elastic-plastic deformation to critical double impulse using energy balance and quadratic approximation of damping force-deformation relation: (a) CASE 1: elastic range, (b) CASE 2: yielding after 2nd impulse, (c) CASE 3: yielding after 1st impulse (●: 1st impulse, ▲: 2nd impulse) (Kojima et al. 2017, 2018).

5.6.2 CASE 1: Elastic response even after second impulse

First of all, consider CASE 1, which is the case of the elastic response, even after the second impulse. Figure 5.8(a) shows the process for evaluating the maximum deformations u_{max1}, u_{max2} after the first impulse and the second impulse, respectively, for the elastic case (CASE 1). Although the exact solution for the elastic response of the system with viscous damping was derived in Section 5.5, an approximate closed-form expression of the maximum deformation is derived here by using the quadratic function approximation for the damping force-deformation relation in terms of the maximum deformation. The approximate expression that is derived in this section is a good approximation of the exact expression that was obtained in Section 5.5.

The work done by the damping force after the first impulse is derived by using the quadratic function approximation in terms of the maximum deformation for the damping force-deformation relation. The damping force-deformation relation after the first impulse is to be approximated by a quadratic function with vertex $(u, f_D) = (-u_{max1}, 0)$ and passing through the point $(u, f_D) = (0, -cV)$. Then, f_D can be obtained as

$$f_D = -cV\sqrt{1 + (u/u_{max1})} \qquad (5.13)$$

The work done by the damping force in this process can be obtained by integrating Eq. (5.13) from $u = 0$ to $u = -u_{max1}$.

$$\int_0^{-u_{max1}} f_D du = \int_0^{-u_{max1}} \left\{ -cV\sqrt{1 + (u/u_{max1})} \right\} du = (2/3) cVu_{max1} \quad (5.14)$$

The energy balance law between the point of action of the first impulse and the point at which the maximum deformation is attained can be expressed as follows by using Eq. (5.14).

$$mV^2/2 = ku_{max1}^2/2 + (2/3)cVu_{max1} \qquad (5.15)$$

From Eq. (5.15), the maximum deformation u_{max1} can be obtained by

$$u_{max1}/d_y = \left\{ -(4/3)h + \sqrt{(16/9)h^2 + 1} \right\}(V/V_y). \qquad (5.16)$$

Similarly, the maximum deformation u_{max2} after the second impulse can be derived. The velocity v_c at the zero-restoring-force timing in the first unloading process can be obtained as follows by using the critical time interval $t_0^c = T_1'/2$ and Eq. (5.7b).

$$v_c = V\exp\left(-\pi h/\sqrt{1 - h^2}\right) \qquad (5.17)$$

The work done by the damping force is derived by using the quadratic function approximation as in the case after the first impulse. The damping force-deformation relation after the second impulse is approximated by a quadratic function with vertex $(u, f_D) = (u_{max2}, 0)$ and passing through the point $(u, f_D) = (0, c(v_c + V))$, as shown in Figure 5.8(a). The damping force f_D after the second impulse can be obtained as follows.

$$f_D = c(v_c + V)\sqrt{1 - (u/u_{max2})} \tag{5.18}$$

The work done by the damping force after the second impulse can be obtained by integrating Eq. (5.18) from $u = 0$ to $u = u_{max2}$.

$$\int_0^{u_{max2}} f_D du = \int_0^{u_{max2}} \left\{ c(v_c + V)\sqrt{1 - (u/u_{max2})} \right\} du = (2/3)c(v_c + V)u_{max2} \tag{5.19}$$

The energy balance law between the point of action of the second impulse and the point at which the maximum deformation is attained can be expressed as follows by using Eq. (5.19).

$$m(v_c + V)^2 / 2 = k u_{max2}^2 / 2 + (2/3)c(v_c + V)u_{max2} \tag{5.20}$$

From Eqs. (5.17) and (5.20), u_{max2} can be obtained by

$$u_{max2}/d_y = \left(1 + e^{-\pi h/\sqrt{1-h^2}}\right)\left\{-(4/3)h + \sqrt{(16/9)h^2 + 1}\right\}(V/V_y) \tag{5.21}$$

5.6.3 CASE 2: Plastic deformation only after the second impulse

Secondly, consider CASE 2, where the EPP SDOF system with viscous damping enters the yielding stage only after the second impulse. Figure 5.8(b) shows the process for evaluating the maximum deformations u_{max1}, u_{max2} after the first impulse and the second impulse, respectively, in CASE 2. If the maximum deformation u_{max2} after the second impulse attains the yield deformation d_y, the system enters the plastic range after the second impulse for the input larger than this boundary. Therefore, the boundary input velocity level between CASE 1 and CASE 2 can be obtained as follows from Eq. (5.21) and $u_{max2} = d_y$.

$$V/V_y = \left(1 + e^{-\pi h/\sqrt{1-h^2}}\right)^{-1}\left\{(4/3)h + \sqrt{(16/9)h^2 + 1}\right\} \tag{5.22}$$

Since the maximum deformation just after the first impulse is in the elastic range, the maximum deformation u_{max1} after the first impulse in

CASE 2 is also obtained by Eq. (5.16). The maximum deformation $u_{\max2}$ after the second impulse in CASE 2 is derived in this section. The work done by the damping force in CASE 2 is obtained by using the quadratic function approximation. As in CASE 1, v_c in CASE 2 can be expressed by Eq. (5.17) due to the elastic response just after the first impulse. The work done by the damping force after the second impulse in CASE 2 can be expressed by Eq. (5.19) by using the quadratic function approximation, as in CASE 1. The energy balance law between the point of action of the second impulse and the point at which the maximum deformation is attained can be expressed as follows by using Eq. (5.19).

$$m\left(v_c + V\right)^2 / 2 = f_y d_y / 2 + f_y \left(u_{\max2} - d_y\right) + \left(2/3\right)c\left(v_c + V\right)u_{\max2} \quad (5.23)$$

From Eqs. (5.17) and (5.23), the maximum deformation $u_{\max2}$ after the second impulse can be obtained as

$$\frac{u_{\max2}}{d_y} = \frac{1 + \left\{1 + \exp\left(-\pi h/\sqrt{1-h^2}\right)\right\}^2 \left(V/V_y\right)^2}{2 + \left(8h/3\right)\left\{1 + \exp\left(-\pi h/\sqrt{1-h^2}\right)\right\}\left(V/V_y\right)}. \quad (5.24)$$

5.6.4 CASE 3: Plastic deformation, even after the first impulse

Finally, consider CASE 3, where the EPP SDOF system with viscous damping enters the yielding stage, even after the first impulse. Figure 5.8(c) shows the process for evaluating the maximum deformations $u_{\max1}$, $u_{\max2}$ after the first impulse and the second impulse, respectively, in CASE 3. If the maximum deformation $u_{\max1}$ after the first impulse attains the yield deformation d_y, the system enters the plastic range after the first impulse for the input larger than this boundary. Therefore, the boundary input velocity level between CASE 2 and CASE 3 can be obtained as follows from Eq. (5.16) and $u_{\max1} = d_y$.

$$V/V_y = \left(4/3\right)h + \sqrt{\left(16/9\right)h^2 + 1} \quad (5.25)$$

The maximum deformation $u_{\max1}$ after the first impulse is derived in the next step. The work done by the damping force after the first impulse in CASE 3 can be expressed by Eq. (5.14) by using the quadratic function approximation in terms of the maximum deformation, as in CASE 1. The energy balance law between the point of action of the first impulse and the point at which the maximum deformation is attained can be expressed as follows by using Eq. (5.14).

$$mV^2/2 = f_y d_y/2 + f_y\left(u_{\text{max}1} - d_y\right) + (2/3)cVu_{\text{max}1} \qquad (5.26)$$

From Eq. (5.26), the maximum deformation $u_{\text{max}1}$ after the first impulse can be obtained by

$$u_{\text{max}1}/d_y = \left\{(V/V_y)^2 + 1\right\}/\left\{2 + (8h/3)(V/V_y)\right\}. \qquad (5.27)$$

The maximum deformation $u_{\text{max}2}$ after the second impulse is derived in the next step. The velocity v_c at the zero-restoring-force timing after the first impulse can be obtained as follows by solving the equation of motion in the unloading process.

$$v_c = V_y \exp\left[\left(-h/\sqrt{1-h^2}\right)\left\{0.5\pi + \arctan\left(h/\sqrt{1-h^2}\right)\right\}\right] \qquad (5.28)$$

The detailed derivation of Eq. (5.28) is shown in Appendix 2. The work done by the damping force is derived by using the quadratic function approximation in terms of the maximum deformation. The damping force-deformation relation after the second impulse is approximated by a quadratic function with vertex $(u, f_D) = (u_{\text{max}2}, 0)$ and passing the point $(u, f_D) = (-u_{\text{max}1} + d_y, c(v_c + V))$, as shown in Figure 5.8(c). The damping force f_D can then be obtained as follows.

$$f_D = c(v_c + V)\sqrt{(u_{\text{max}2} - u)/(u_{\text{max}1} + u_{\text{max}2} - d_y)} \qquad (5.29)$$

The work done by the damping force after the second impulse can be obtained by integrating Eq. (5.29) from $u = -u_{\text{max}1} + d_y$ to $u = u_{\text{max}2}$.

$$\int_{-u_{\text{max}1}+d_y}^{u_{\text{max}2}} \left\{c(v_c + V)\sqrt{(u_{\text{max}2} - u)/(u_{\text{max}1} + u_{\text{max}2} - d_y)}\right\} du$$
$$= (2/3)c(v_c + V)(u_{\text{max}1} + u_{\text{max}2} - d_y) \qquad (5.30)$$

The energy balance law between the point of action of the second impulse and the point at which the maximum deformation is attained can be expressed as follows by using Eq. (5.30).

$$m(v_c + V)^2/2 = f_y d_y/2 + f_y\left(u_{\text{max}1} + u_{\text{max}2} - 2d_y\right)$$
$$+ (2/3)c(v_c + V)(u_{\text{max}1} + u_{\text{max}2} - d_y) \qquad (5.31)$$

From Eqs. (5.28) and (5.31), u_{max2} can be obtained as

$$\frac{u_{max2}}{d_y} = -\frac{u_{max1}}{d_y} + 1 + \frac{\left[\frac{V}{V_y} + \exp\left\{\frac{-h}{\sqrt{1-h^2}}\left(\frac{\pi}{2} + \arctan\frac{h}{\sqrt{1-h^2}}\right)\right\}\right]^2 + 1}{2 + \frac{8h}{3}\left[\frac{V}{V_y} + \exp\left\{\frac{-h}{\sqrt{1-h^2}}\left(\frac{\pi}{2} + \arctan\frac{h}{\sqrt{1-h^2}}\right)\right\}\right]} \quad (5.32)$$

5.6.5 Maximum deformation under the critical double impulse with respect to the input velocity level

Figure 5.9 shows a comparison of an approximate expression for the maximum deformation $u_{max}/d_y = \max(u_{max1}/d_y, u_{max2}/d_y)$ under the critical double impulse with respect to the input velocity level V/V_y, with the maximum deformation calculated by the nonlinear time-history response analysis without the quadratic function approximation of the damping force-deformation relation. The Newmark-β method (the constant average acceleration method with the time increment $\Delta t/T_1 = 10^{-4}$) is used in the nonlinear time-history response analysis, and the maximum deformation under the critical double impulse for various time intervals is calculated by changing the impulse time interval t_0 in a parametric manner. The damping ratios $h = 0, 0.02, 0.05, 0.1, 0.2, 0.5$ are employed in Figure 5.9. The approximate expression for $h = 0$ is equal to the closed-form expression of the undamped system that was derived in Kojima and Takewaki (2015a) and Chapter 2.

It can be observed from Figure 5.9 that, in comparison with the critical response obtained by the time-history response analysis, the approximate closed-form expression that was derived in Sections 5.6.2, 5.6.3, and 5.6.4 can simulate the elastic-plastic response of the EPP SDOF system with viscous damping under the critical double impulse with reasonable accuracy, except when $V/V_y > 3$ in the model with $h = 0.2, 0.5$. The region in which $u_{max2} > u_{max1}$ is satisfied decreases as the damping ratio increases.

Figure 5.10 shows the normalized critical time interval $t_0{}^c$ that was calculated by the time-history response analysis with respect to the input velocity level. The closed-form expression of the critical time interval $t_0{}^c$ of the undamped EPP SDOF system was derived by solving the equation of motion directly in Kojima and Takewaki (2015a) and Chapter 2. Since it is difficult to derive the critical time interval for the EPP SDOF system with viscous damping by solving the equation of motion, the critical time interval is obtained here by using the time-history response analysis. From Figure 5.10, the time interval becomes shorter as the damping ratio increases at the input velocity level for which the system enters the plastic region just after the first impulse (CASE 3). This is because the plastic deformation after the first impulse for the specific input level V/V_y becomes smaller as the damping

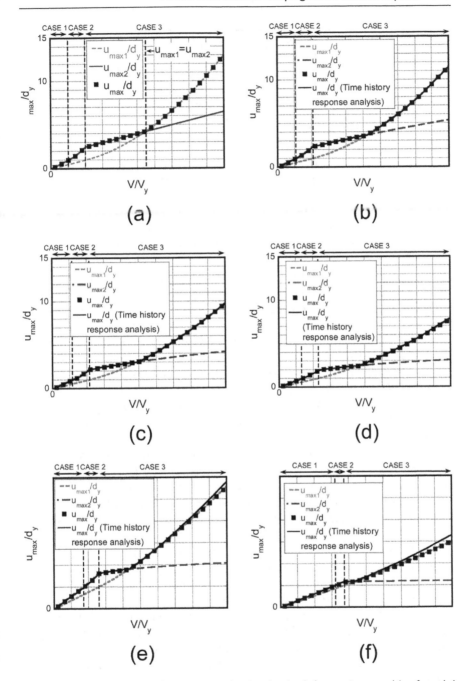

Figure 5.9 Comparison of maximum elastic-plastic deformation u_{max}/d_y of model with viscous damping under critical double impulse using quadratic approximation of damping force-deformation relation with that obtained by time-history response analysis: (a) $h = 0$, (b) $h = 0.02$, (c) $h = 0.05$, (d) $h = 0.1$, (e) $h = 0.2$, (f) $h = 0.5$ (Kojima et al. 2017, 2018).

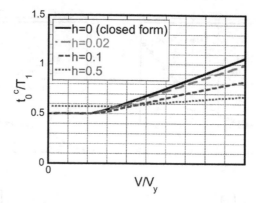

Figure 5.10 Critical impulse timing $t_0{}^c/T_1$ for varied input level V/V_y in models with various damping ratios (Kojima et al. 2017, 2018).

ratio increases. In contrast to CASES 1 and 2, the time interval of impulses becomes larger as the damping ratio increases, because of $t_0{}^c = T_1'/2$.

Figure 5.11 shows a comparison of the restoring force-deformation relation and the damping force-deformation relation that were obtained by using the quadratic function approximation with those that were obtained by the time-history response analysis. The damping ratio $h = 0.05$ was employed in this analysis. Figures 5.11(a), (b), and (c) present the comparisons for $V/V_y = 0.4$ in CASE 1, $V/V_y = 0.8$ in CASE 2, and $V/V_y = 2.0$ in CASE 3, respectively. According to Figure 5.11, the damping force-deformation relation can be approximated properly by using a quadratic function.

Figure 5.12 shows the normalized maximum deformation $u_{max2}/u_{max2}{}^c$ after the second impulse with respect to the varying time interval t_0 and the restoring force at the time $t = t_0$ after the first impulse for $V/V_y = 2.0$, $h = 0.05$, obtained by the time-history response analysis. In Figure 5.12, u_{max2} and $u_{max2}{}^c$ denote the maximum deformations after the second impulse under the double impulse for the varying time interval and that under the critical double impulse (the maximum value of u_{max2}), respectively. The maximum deformations u_{max2} and $u_{max2}{}^c$ are calculated by the time-history response analysis. One of the ordinates of Figure 5.12 denotes u_{max2}, normalized by $u_{max2}{}^c$, and the other denotes the restoring force f at $t = t_0$, normalized by the yield force f_y. According to Figure 5.12, the restoring force becomes zero at the time $t = t_0$ at which u_{max2} reaches $u_{max2}{}^c$. Therefore, the zero-restoring-force timing after the first impulse is the critical timing of the second impulse of the EPP SDOF system with viscous damping.

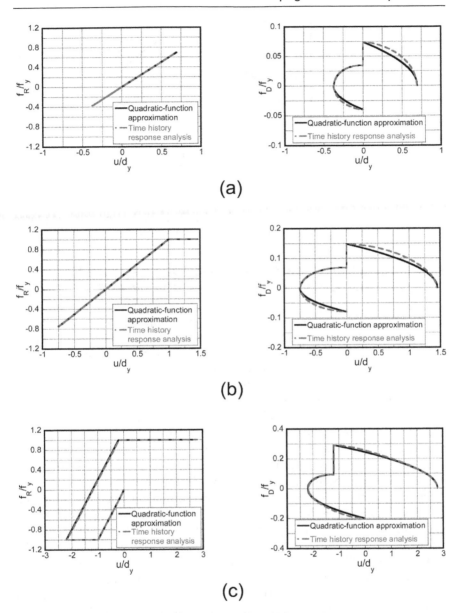

Figure 5.11 Comparison of elastic-plastic responses to critical double impulse obtained using quadratic approximation of damping force-deformation relation with those obtained by time-history response analysis (h = 0.05): (a) V/V_y = 0.4, (b) V/V_y = 0.8, (c) V/V_y = 2.0 (Kojima et al. 2017, 2018).

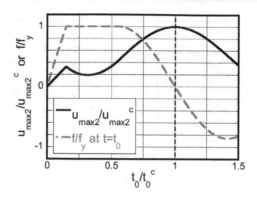

Figure 5.12 Maximum deformation u_{max2}/d_y after 2nd impulse and restoring force f/f_y at $t = t_0$ for t_0/t_0^c $(V/V_y = 2.0, h = 0.05)$ (Kojima et al. 2017, 2018).

5.7 ACCURACY CHECK BY TIME-HISTORY RESPONSE ANALYSIS TO ONE-CYCLE SINUSOIDAL WAVE

To investigate the accuracy of using the double impulse as a substitute for the one-cycle sinusoidal wave in representing the fling-step near-fault ground motions, time-history response analysis of the EPP SDOF system with viscous damping under the one-cycle sinusoidal wave was conducted. The maximum velocity V_p of the corresponding one-cycle sinusoidal wave is adjusted so that the maximum Fourier amplitude of the one-cycle sinusoidal wave is equal to that of the double impulse (Kojima and Takewaki 2015a, 2016a, b). The adjustment procedure can be found in Section 1.2 in Chapter 1. Eq. (5.2) is used as an acceleration waveform of the one-cycle sinusoidal wave. The period T_p of the one-cycle sinusoidal wave is $T_p = 2t_0^c$, where t_0^c is calculated by time-history response analysis, as shown in Figure 5.10.

Figure 5.13 shows a comparison of the maximum deformation of the EPP SDOF system with viscous damping under the critical double impulse with that under the corresponding one-cycle sinusoidal wave for the damping ratios $h = 0, 0.02, 0.05, 0.1, 0.2, 0.5$. The maximum deformation of the undamped EPP SDOF system under the critical double impulse is in good agreement with that under the one-cycle wave in the range of the input velocity level $V/V_y < 3$. As the damping ratio increases, the maximum deformation under the critical double impulse corresponds well to that under the one-cycle sinusoidal wave in a wider range of input velocity level. This is because the maximum deformation after the first impulse (governing the maximum deformation in the input velocity range $V/V_y > 3$) exhibits better correspondence with that under the first half-cycle of the corresponding one-cycle sinusoidal wave as the damping ratio increases. This result clearly indicates that the adjustment procedure of the input level of the double impulse and the corresponding one-cycle sinusoidal wave based on the

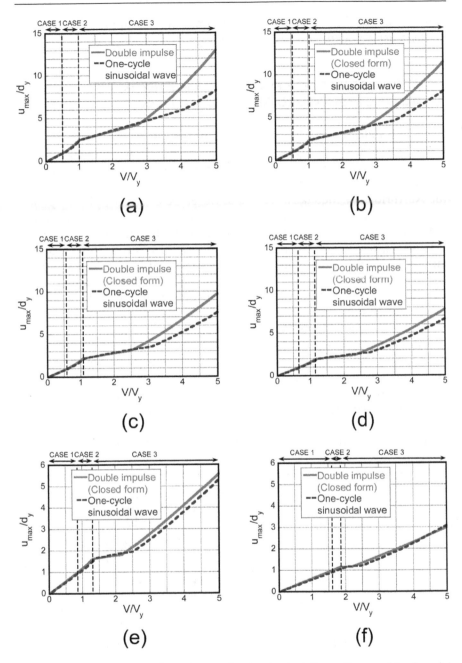

Figure 5.13 Comparison of maximum elastic-plastic deformation u_{max}/d_y of model with viscous damping under critical double impulse (quadratic-function approximation) with that under equivalent one-cycle sine wave: (a) $h = 0$, (b) $h = 0.02$, (c) $h = 0.05$, (d) $h = 0.1$, (e) $h = 0.2$, (f) $h = 0.5$ (Kojima et al. 2017, 2018).

equivalence of the maximum Fourier amplitude is appropriate for the elastic-plastic system with viscous damping.

5.8 APPLICABILITY OF PROPOSED THEORY TO ACTUAL RECORDED GROUND MOTION

The applicability of the theory explained in this chapter to actual recorded ground motions is investigated through the comparison of the critical elastic-plastic response under the near-fault ground motion with that under the critical double impulse. The Rinaldi station fault-normal component during the Northridge earthquake in 1994 and the Kobe University NS component (almost fault-normal) during the Hyogoken-Nanbu (Kobe) earthquake in 1995 were used as the near-fault ground motions. The accelerograms of these two ground motions are shown in Figure 5.14. Although these are the fault-normal ground motions, these are represented by the double impulse in this section. The main part of the recorded ground motion acceleration is modeled as a one-cycle sinusoidal wave, as shown in Figure 5.14, and the one-cycle sinusoidal wave is substituted by the double impulse by using the method shown in Sections 5.2 and 1.2 (Eq. (1.16)).

Although the critical double impulse was determined for a given structural parameter V_y in Section 5.6, the structural parameter was selected to approximately maximize the response for a given input velocity V of the actual recorded ground motion in this section. This procedure is similar to the elastic-plastic response spectrum (changing the strength parameter), which was developed in 1960–1970 (Veletsos et al. 1965). In this section, a method for evaluating the critical elastic-plastic response under the

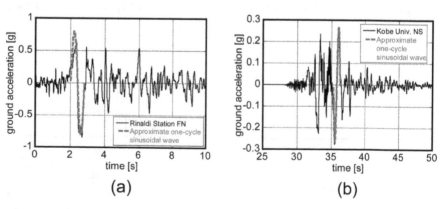

Figure 5.14 Recorded near-fault ground motion and corresponding one-cycle sine wave: (a) Rinaldi Station FN comp. (Northridge 1994), (b) Kobe Univ. NS comp. (Hyogoken-Nanbu 1995) (Kojima et al. 2017, 2018).

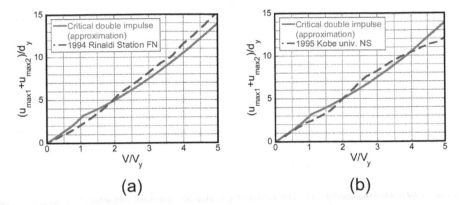

Figure 5.15 Comparison of maximum elastic-plastic deformation (double amplitude) of model with viscous damping ($h = 0.05$) under critical double
impulse (quadratic-function approximation) and recorded ground
motions: (a) Rinaldi Sta. FN comp., (b) Kobe Univ. NS comp. (Kojima
et al. 2018).

near-fault ground motion that is used in the literature (Kojima and Takewaki
2016b) was employed. The input velocity level of the Rinaldi station fault-
normal component is $V = 1.64$[m/s] and that of the Kobe University NS
component is $V = 0.677$[m/s] by the method in the literature (Kojima and
Takewaki 2016b).

Figures 5.15(a), (b) show a comparison of the critical elastic-plastic
response of the EPP SDOF system with viscous damping under the Rinaldi
station fault-normal component and the Kobe University NS component
with the proposed closed-form expression of the elastic-plastic response
under the critical double impulse. The ordinate presents the maximum
amplitude of deformation (the sum of u_{max1} and u_{max2}), and the abscissa is
the input velocity level V/V_y. In comparison with the undamped case shown
in Figures 2.13(a), (b) in Chapter 2 (Kojima and Takewaki 2016a, b), the
elastic-plastic response under the critical double impulse corresponds well
to the critical elastic-plastic response under the actual recorded ground
motion in a wide range of the input velocity level, owing to the existence of
viscous damping.

5.9 SUMMARIES

The double impulse was introduced as a good substitute for the one-cycle
sinusoidal wave in representing the main part of a near-fault ground motion.
A closed-form expression was derived for the maximum deformation of the
elastic–perfectly plastic (EPP) single-degree-of-freedom (SDOF) system with

viscous damping under the critical double impulse. The summaries are as follows:

1. A closed-form expression was derived approximately for the maximum deformation of the EPP SDOF system with viscous damping under the critical double impulse. Since it seems difficult to derive the exact response of the EPP SDOF system with viscous damping by solving the differential equation, even by using the double impulse, an approximate but explicit expression was obtained. It uses (1) a quadratic function to approximate the damping force-deformation relation, (2) the assumption that the zero restoring-force timing in the unloading process is the critical timing of the second impulse, and (3) the energy balance law for the elastic-plastic system with viscous damping.

2. The accuracy of the quadratic function approximation of the damping force-deformation relation and the validity of the assumption on the critical timing were investigated by using nonlinear time-history response analysis. It was demonstrated that the damping force-deformation relation can be approximated by a quadratic function with reasonable accuracy. It was also confirmed that the zero-restoring-force timing is the critical timing of the second impulse, which maximizes the peak deformation after the second impulse.

3. The validity of using the double impulse as a substitute for the near-fault ground motion was investigated through the comparison with the elastic-plastic response under the corresponding one-cycle sinusoidal wave. Although the accuracy was not necessarily sufficient for the undamped system compared to the corresponding one-cycle sinusoidal wave in the range of the larger input velocity level $V/V_y > 3$ (Kojima and Takewaki 2015a and Chapter 2), the maximum deformation of the system with larger viscous damping under the critical double impulse corresponds well to that under the corresponding one-cycle sinusoidal wave, even in the range of the larger input velocity level.

4. The applicability of the present method using the critical double impulse (defined in Section 5.6) to actual earthquake ground motions was investigated through comparison with the critical elastic-plastic response under actual recorded ground motions. It was demonstrated that the accuracy is sufficient in the wider range of the input velocity level.

In this chapter, an approximate but explicit critical response of a damped EPP SDOF system under the double impulse was derived. It is expected that the formulation is applied to the critical response under the triple impulse explained in Chapter 3 (Kojima and Takewaki 2015b).

REFERENCES

Kojima, K., and Takewaki, I. (2015a). Critical earthquake response of elastic-plastic structures under near-fault ground motions (Part 1: Fling-step input), *Frontiers in Built Environment*, 1: 12.

Kojima, K., and Takewaki, I. (2015b). Critical earthquake response of elastic-plastic structures under near-fault ground motions (Part 2: Forward-directivity input), *Frontiers in Built Environment*, 1: 13.

Kojima, K., and Takewaki, I. (2016a). Closed-form critical earthquake response of elastic-plastic structures on compliant ground under near-fault ground motions, *Frontiers in Built Environment*, 2: 1.

Kojima, K., and Takewaki, I. (2016b). Closed-form critical earthquake response of elastic-plastic structures with bilinear hysteresis under near-fault ground motions, *J. Struct. Constr. Eng., AIJ*, 81(726): 1209–1219 (in Japanese).

Kojima, K., Saotome, Y., and Takewaki, I. (2017, 2018). Critical earthquake response of an SDOF elastic-perfectly plastic model with viscous damping under double impulse as a substitute for near-fault ground motion, *Japan Architectural Review (Int. J. of Japan Architectural Review for Engineering and Design)*, Wiley, English Version, Vol. 1, Issue 2, pp.207–220, 2018). (Japanese version, *J. Struct. and Construction Eng.*, No.735, pp. 643–652, 2017).

Takewaki, I. (2007). *Critical excitation methods in earthquake engineering*, Elsevier, Amsterdam 2007, Second edition in 2013, London.

Veletsos, A. S., Newmark, N. M., and Chelapati, C. V. (1965). Deformation spectra for elastic and elasto-plastic systems subjected to ground shock and earthquake motions, *Proc. of the Third World Conference on Earthquake Engineering, New Zealand*, Vol. II, pp. 663–682.

Caughey, T. K. (1960). Sinusoidal excitation of a system with bilinear hysteresis, *J. Appl. Mech.*, 27(4): 640–643.

Iwan, W. D. (1961). The dynamic response of bilinear hysteretic systems, Ph.D. Thesis, California Institute of Technology.

APPENDIX 1: CRITICAL IMPULSE TIMING FOR LINEAR ELASTIC SYSTEM WITH VISCOUS DAMPING

The critical timing of the second impulse for a linear elastic system with viscous damping is explained. The maximum deformation u_{max2} after the second impulse for a variable impulse time interval can be obtained by Eq. (5.10b). The maximum deformation u_{max2} after the second impulse is maximized at the time interval at which the derivative of Eq. (5.10b) with respect to the impulse time interval t_0 becomes zero. From Figure 5.5, the maximum value of u_{max2} decreases due to the effect of viscous damping as the impulse time interval becomes longer. Therefore, the maximum value of u_{max2} is investigated in the range $0 < t_0 < T_1'$. However, it is complicated to obtain the derivative of Eq. (5.10b) with respect to the impulse time interval t_0. In a simple manner, $t_0 = T_1'/2$ was substituted into the equation of du_{max2}/dt_0. It was confirmed that du_{max2}/dt_0 is zero at that value. The parameters t_{max2}, ϕ, θ in Eq. (5.10b) can be obtained from Eqs. (5.9a, b) and Eq. (5.11b), respectively. It is noted that t_{max2} and θ are functions of t_0.

APPENDIX 2: VELOCITY AT ZERO RESTORING FORCE AFTER ATTAINING u_{max1} IN CASE 3

The velocity of mass at the zero-restoring-force timing after the first impulse in CASE 3 can be obtained by solving the equation of motion in the unloading process. An outline of the derivation of v_c is shown here. The restoring-force characteristic can be described by using the maximum deformation u_{max1} after the first impulse.

$$f_R = ku + k\left(u_{max1} - d_y\right),$$ (5.A1)

where an approximate expression of u_{max1} is obtained from Eq. (5.27). The equation of motion (free vibration) in the unloading process can be expressed by using Eq. (5.A1).

$$m\ddot{u} + c\dot{u} + ku + k\left(u_{max1} - d_y\right) = 0$$ (5.A2)

From Eq. (5.A2), the deformation and velocity in the unloading process can be described by

$$u = -\left(1/\sqrt{1-h^2}\right)d_y e^{-h\omega_1 t}\cos\left(\omega_1' t - \phi\right) - \left(u_{max1} - d_y\right),$$ (5.A3)

$$\dot{u} = \left(1/\sqrt{1-h^2}\right)V_y e^{-h\omega_1 t}\sin\left(\omega_1' t\right),$$ (5.A4)

where the starting time of the unloading process (the point $-u_{max1}$ in Figure 5.8(c)) is taken as $t = 0$. The parameter ϕ is obtained from Eq. (5.9a). From Eqs. (5.A1) and (5.A3), the time t^c at which the restoring-force becomes zero can be obtained as follows.

$$t^c = \left[0.25 + \left\{\arctan\left(h/\sqrt{1-h^2}\right)\right\}/(2\pi)\right]T_1'$$ (5.A5)

The velocity v_c at the zero restoring force can then be obtained by substituting Eq. (5.A5) into Eq. (5.A4).

$$v_c = V_y \exp\left[\left(-h/\sqrt{1-h^2}\right)\left\{0.5\pi + \arctan\left(h/\sqrt{1-h^2}\right)\right\}\right]$$ (5.A6)

Chapter 6

Critical steady-state response of a bilinear hysteretic SDOF model under multi-impulse

6.1 INTRODUCTION

A set of multiple impulses is used in this chapter as in Chapter 4 to substitute many-cycle harmonic waves that represent the long-duration earthquake ground motion. Figure 6.1 shows an actual example of the resonant response recorded in a high-rise building in Osaka, Japan during the 2011 off the Pacific coast of Tohoku earthquake (Takewaki et al. 2011, 2013). Although damage only to nonstructural components was observed in this building, the development of damage to structural components in future's bigger ones should be taken into account from the viewpoint of resilience. This actual incident clearly implies the warning to consider the response under long-duration ground motion carefully.

A closed-form expression is derived here for the elastic-plastic response of a single-degree-of-freedom (SDOF) model with bilinear hysteresis under the "critical multiple impulse input." As in the case of elastic–perfectly plastic models, an advantageous feature can be used such that only free-vibration occurs under the multiple impulse and the energy balance approach plays a key role in the derivation of the closed-form expression on complicated maximum elastic-plastic response. It is demonstrated that the critical value of inelastic maximum deformation and the corresponding critical impulse timing can be obtained explicitly depending on the input level.

An undamped bilinear hysteretic SDOF system used in this chapter is explained in Section 6.2. The closed-form expressions are derived for the elastic-plastic steady-state responses under the critical multi-impulse and for the critical time intervals of two cases in Section 6.3. CASE 1 is the case where each impulse acts at the zero-restoring-force timing in the unloading process, and the other case, CASE 2, is the case where each impulse acts at the zero-restoring-force timing in the loading process. It is investigated whether the response under the multi-impulse with the critical time interval obtained in Section 6.3 converges to the steady state in which each impulse acts at the zero-restoring-force point in Section 6.4. The accuracy of using the multi-impulse as a substitute of the long-duration ground motion is checked through the comparison with the response under the corresponding

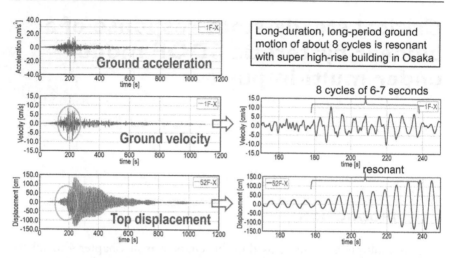

Figure 6.1   Resonant response of super high-rise building in Osaka, Japan during the 2011 off the Pacific coast of Tohoku earthquake under long-duration, long-period ground motion (Takewaki et al. 2013).

multi-cycle sinusoidal wave in Section 6.5. The validity of the critical time interval obtained in Section 6.3 is confirmed by time-history response analysis of the bilinear hysteretic SDOF system under multi-impulse with various impulse time intervals in Section 6.6. The applicability of the critical impulse timing obtained in Section 6.3 to the corresponding sinusoidal wave is investigated in Section 6.7. The accuracy of the proposed closed-form steady-state response under the critical multi-impulse is also investigated in Section 6.8 through the comparison with the resonance curve under the sinusoidal wave provided by Iwan (1961). The conclusions are summarized in Section 6.9.

6.2 BILINEAR HYSTERETIC SDOF SYSTEM

Consider an undamped bilinear hysteretic SDOF system of mass m and stiffness k subjected to the multi-impulse with the equal time interval as shown in Figures 6.2(a), (b). V is the given initial velocity (the input velocity level of each impulse) and t_0 is the equal time interval between two consecutive impulses. The ratio of the post-yield stiffness to the initial elastic stiffness is expressed by α. In this chapter, the case of $\alpha > 0$ is treated. The yield deformation and the yield force are denoted by d_y and f_y. Let $\omega_1 = \sqrt{k/m}$, u, and f denote the undamped natural circular frequency, the displacement of the mass relative to the ground (deformation of the system), and the restoring force of the model, respectively. As in the previous chapters, the time derivative is denoted by an over-dot. In Section 6.3, these parameters will be treated as normalized ones to capture the intrinsic relation between the input parameters and the elastic-plastic response. However, numerical

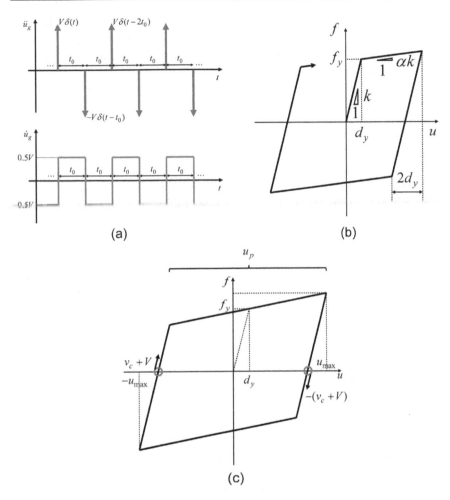

(a) (b)

(c)

Figure 6.2 Impulse input and bilinear hysteretic restoring force-deformation characteristic: (a) multi-impulse with equal time interval t_0, (b) bilinear hysteretic restoring-force characteristic, (c) steady-state loop under critical multi-impulse (Kojima and Takewaki 2017).

investigations will be made in Sections 6.4 to 6.8 to demonstrate an example of models with actual parameters.

6.3 CLOSED-FORM EXPRESSION FOR ELASTIC-PLASTIC STEADY-STATE RESPONSE TO CRITICAL MULTI-IMPULSE

In Kojima and Takewaki (2015a, b, c), some closed-form expressions for the critical elastic-plastic responses of an elastic–perfectly plastic (EPP) SDOF system under the double, triple, and multi-impulses were derived (Chapters 2–4 in this book). A closed-form expression of the maximum deformation of a bilinear hysteretic SDOF system under the double impulse was also derived in the reference (Kojima and Takewaki 2016). In this chapter, a closed-form expression of the steady-state elastic-plastic response of a bilinear hysteretic SDOF system under the critical multi-impulse is derived and explained in detail.

First of all, it is important to understand that there exists a steady state with an elastic-plastic response for a bilinear hysteretic (positive second stiffness) SDOF system under the multi-impulse. In the 1960s, the existence itself was not certain before a laborious computational task was completed with limited computational resources (Caughey 1960a, b, Iwan 1961, 1965a, b).

The elastic-plastic response after each impulse input can be expressed by the instantaneous change of velocity of the mass by V and only free vibration appears after each impulse input. Since the elastic-plastic response of the bilinear hysteretic SDOF system under the multi-impulse can be expressed by the continuation of free vibrations, the plastic deformation amplitude and the maximum deformation can be derived by an energy balance approach without solving the equation of motion directly. The kinetic energy introduced at the input time of each impulse is transformed into the combination of the hysteretic energy and the strain energy (see Chapter 4 for EPP SDOF systems). Each impulse's critical timing corresponds to the phase with the zero restoring force (no strain energy) and kinetic energy alone appears in this phase as mechanical energies. By using this rule, the maximum deformation can be evaluated in a simple manner. In Kojima and Takewaki (2015c) and Chapter 4, the closed-form expression for plastic deformation amplitude and the critical timing of the EPP SDOF system under the critical multi-impulse were derived. To derive the closed-form plastic deformation amplitude and critical timing, a modified multi-impulse, in which the first and second impulses are modified so that the second impulse is given at the zero restoring force, was introduced in the reference (Kojima and Takewaki 2015c) and Chapter 4. However, the elastic-plastic response of the present bilinear hysteretic SDOF system with $\alpha > 0$ cannot become stable under the first few impulses even in the condition that each impulse acts at the zero restoring force and the response converges to a steady state as shown in Figure 6.2(c) after a sufficiently large number of repetitive impulses. In this section, the steady state in which each impulse acts at the zero-restoring-force point is assumed and the closed-form expressions for the elastic-plastic response and the critical timing are derived by using the assumption of the existence of the steady state and the energy balance approach. The convergence of the response under the multi-impulse with the equal time interval obtained in Section 6.3.4 into the steady state will be verified in Section 6.4. The convergence of the response under a harmonic wave into the

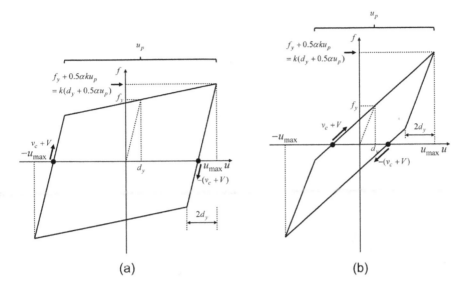

Figure 6.3 Restoring force-deformation relation under critical multi-impulse: (a) CASE 1 ($u_p/d_y \leq 2/\alpha$): impulse in unloading process; (b) CASE 2 ($u_p/d_y > 2/\alpha$): impulse in loading process (Kojima and Takewaki 2017).

steady state was also confirmed in the milestone thesis in the early stage (Iwan 1961).

The steady state under the critical multi-impulse can be classified into two cases depending on the plastic deformation level as shown in Figures 6.3(a), (b). Figures 6.3(a), (b) show the case (CASE 1) in which each impulse acts at the zero-restoring-force timing in the unloading process and the case (CASE 2) in which each impulse acts at the zero-restoring-force timing in the loading process, respectively. This kind of classification was not necessary for the EPP SDOF system (Kojima and Takewaki 2015c and Chapter 4). The boundary between CASE 1 and CASE 2 is given by $u_p/d_y = 2/\alpha$, and this condition will be derived in Section 6.3.1.

6.3.1 CASE 1: Impulse in unloading process

First of all, consider CASE 1. The steady-state elastic-plastic response (plastic deformation amplitude and maximum deformation) can be derived for the bilinear hysteretic SDOF system under the critical multi-impulse by using the energy balance law. Figure 6.4(a) shows the schematic diagram for the derivation of the maximum steady-state response in CASE 1 based on the energy balance approach. Figures 6.4(b), (c) show the time histories of the deformation and the restoring force in the steady state with the critical time interval $t_0{}^c$ between two consecutive impulses. The times t_{AB}, t_{BC}, t_{CD} denote the time intervals between points A, B, points B, C, and points C, D,

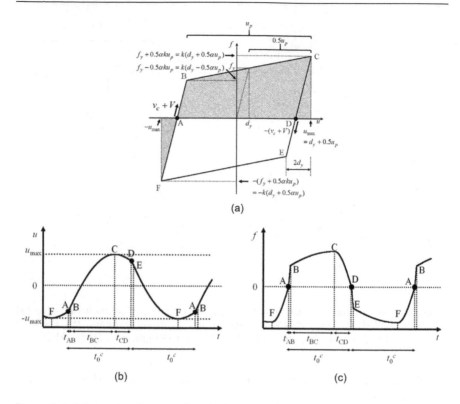

Figure 6.4 Schematic diagram for derivation of maximum deformation under critical multi-impulse based on energy balance approach: (a) restoring force-deformation relation; (b) displacement time history; (c) restoring-force time history (CASE I: $V/V_y < -2 + 2\sqrt{1/\alpha}$) (Kojima and Takewaki 2017).

respectively, in Figure 6.4(a). The closed-form expressions for the time-history responses of the deformation and the restoring force in the steady state with the critical time interval $t_0{}^c (= t_{AB} + t_{BC} + t_{CD})$ between two consecutive impulses can be obtained by solving the equations of motion and substituting the continuation conditions at the transition points (points A, B and C). The closed-form expressions for the time-history responses and the critical time interval are derived in Sections 6.3.4 and Appendix 1.

The velocity v_c at the zero-restoring-force point in the unloading process (point A in Figure 6.4) can be derived by using the energy balance law. The energy balance law between the starting point of unloading (point F in Figure 6.4) and the zero-restoring-force point (point A in Figure 6.4) can be expressed by

$$k\left(d_y + 0.5\alpha u_p\right)^2 / 2 = m v_c^2 / 2 \qquad (6.1)$$

The left-hand side of Eq. (6.1) expresses the elastic strain energy indicated by the triangular shaded area including points A and F in Figure 6.4(a). On the other hand, the right-hand side of Eq. (6.1) presents the kinetic energy at the zero-restoring-force point.

From Eq. (6.1), v_c is expressed in terms of u_p by

$$v_c/V_y = 1 + 0.5\alpha\left(u_p/d_y\right) \tag{6.2}$$

where $V_y = \omega_1 d_y$ is used. V_y denotes the input level of the single impulse at which the SDOF system at rest just attains the yield deformation after the single impulse. This parameter also presents a strength parameter with the velocity dimension.

In the steady state, the plastic deformation u_p after each impulse can be obtained from the energy balance law. The energy balance law between the zero-restoring-force point (point A in Figure 6.4) and the point attaining the maximum deformation (point C in Figure 6.4) can be described by

$$m\left(v_c + V\right)^2/2 = k\left(d_y - 0.5\alpha u_p\right)^2/2 + \alpha k u_p^2/2 + \left(f_y - 0.5\alpha k u_p\right)u_p \tag{6.3}$$

The left-hand side of Eq. (6.3) indicates the kinetic energy computed in terms of the velocity $(v_c + V)$ of mass just after each impulse. On the other hand, the right-hand side of Eq. (6.3) shows the hysteretic and elastic strain energy indicated by the shaded area including points A, B, C in Figure 6.4(a).

Substitution of Eq. (6.2) into Eq. (6.3) and rearrangement of the resulting equation provide the expression for the amplitude of plastic deformation u_p normalized by d_y

$$u_p/d_y = \left\{\left(V/V_y\right)^2 + 2\left(V/V_y\right)\right\}/\left\{2 - 2\alpha - \alpha\left(V/V_y\right)\right\} \tag{6.4}$$

From Eq. (6.4) and Figure 6.4(a), the maximum deformation u_{max} normalized by d_y can be obtained as follows:

$$u_{max}/d_y = 1 + 0.5\left(u_p/d_y\right)$$
$$= 1 + 0.5\left[\left\{\left(V/V_y\right)^2 + 2\left(V/V_y\right)\right\}/\left\{2 - 2\alpha - \alpha\left(V/V_y\right)\right\}\right] \tag{6.5}$$

As the next step, consider the boundary between CASE 1 and CASE 2. In this boundary, the zero-restoring-force point (point A in Figure 6.4) is equal to the point of the yielding initiation (point B in Figure 6.4) and each impulse acts at this point (point A in Figure 6.5(a)). Figures 6.5(a), (b) show the schematic diagram for the derivation of the maximum steady-state response (deformation) in this boundary case based on the energy balance approach.

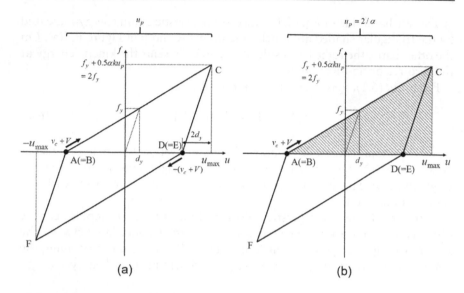

Figure 6.5 Restoring force-deformation relation in boundary case between CASE 1 and CASE 2: (a) Acting points of each impulse; (b) Sum of hysteretic and elastic strain energy after each impulse. (Kojima and Takewaki 2017)

The plastic deformation u_p in this boundary case can be obtained as follows from Figure 6.5(a).

$$f_y + 0.5\alpha k u_p = 2f_y \tag{6.6}$$

From Eq. (6.6), the plastic deformation u_p normalized by d_y in this boundary case can be obtained as follows:

$$u_p/d_y = 2/\alpha \tag{6.7}$$

The boundary input velocity level of the multi-impulse is derived in the next step. From Eq. (6.7) and Figure 6.5(b), the velocity v_c at the zero-restoring-force point (point A in Figure 6.5(a)) can be derived from the energy balance law. The energy balance law between the starting point of unloading (point F in Figure 6.5(b)) and the zero-restoring-force point (point A in Figure 6.5(b)) can be expressed by

$$k\left(2d_y\right)^2/2 = mv_c^2/2 \tag{6.8}$$

The left-hand side of Eq. (6.8) expresses the elastic strain energy at the starting point of unloading. On the other hand, the right-hand side of Eq. (6.8) indicates the kinetic energy at the zero-restoring-force point.

From Eq. (6.8), v_c normalized by V_y can be obtained as follows:

$$v_c/V_y = 2 \tag{6.9}$$

The energy balance law between the zero-restoring-force point (point A in Figure 6.5(b)) and the point attaining the maximum deformation (point C in Figure 6.5(b)) can also be expressed by

$$m\left(v_c + V\right)^2 /2 = \alpha k u_p^{\,2}/2 \tag{6.10}$$

The left-hand side of Eq. (6.10) expresses the kinetic energy computed in terms of the velocity $(v_c + V)$ of mass, which is provided just after each impulse. On the other hand, the right-hand side of Eq. (6.10) indicates the hysteretic and elastic strain energy shown by the shaded area in Figure 6.5(b).

Substitution of Eqs. (6.7) and (6.9) into Eq. (6.10) and rearrangement of the resulting equation provide the boundary input velocity level as follows:

$$V/V_y = -2 + 2\sqrt{1/\alpha} \tag{6.11}$$

6.3.2 CASE 2: Impulse in loading process (second stiffness range)

Consider CASE 2 $(V/V_y > -2 + \sqrt{1/\alpha})$ in the next step. As in CASE 1, the steady-state elastic-plastic response is derived for the bilinear hysteretic SDOF system under the critical multi-impulse by using the energy balance law. Figure 6.6(a) illustrates the maximum steady-state response in CASE 2 based on the energy balance approach. Figures 6.6(b), (c) show the one-cycle time histories of the deformation and the restoring force between two consecutive impulses in the steady state. The times t_{AB}, t_{BC}, t_{CD} denote the time intervals between points A, B, points B, C, and points C, D, respectively, in Figure 6.6(a). The closed-form expressions for the time history responses of the deformation and the restoring force in the steady state with the critical time interval $t_0^c(=t_{AB} + t_{BC} + t_{CD})$ between two consecutive impulses can be obtained by solving the equations of motion and substituting the continuation conditions at the transition points (points A, B and C). The closed-form expressions for the time-history responses and the critical time interval are derived in Section 6.3.4 and Appendix 1.

The velocity v_c at the zero-restoring-force point in the loading process (point A in Figure 6.6(a)) can be derived from the energy balance law. The energy balance law between the starting point of unloading (point E in Figure 6.6) and the zero-restoring-force point (point A in Figure 6.6) can be expressed by

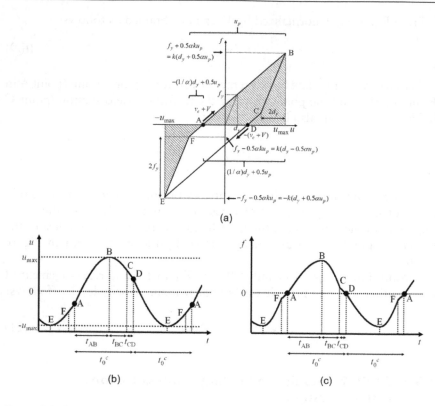

Figure 6.6 Derivation of maximum deformation to critical multi impulse based on energy approach: (a) restoring force-deformation relation; (b) displacement time history; (c) restoring-force time history (CASE 2: $V/V_y \geq -2 + 2\sqrt{1/\alpha}$) (Kojima and Takewaki 2017).

$$k(2d_y)^2 / 2 + (-f_y + 0.5\alpha k u_p)(2d_y)$$
$$+ (\alpha k)\{(-f_y + 0.5\alpha k u_p)/(\alpha k)\}^2 / 2 = m v_c^2 / 2 \qquad (6.12)$$

The left-hand side of Eq. (6.12) shows the elastic strain energy indicated by the shaded area including points A, F, E in Figure 6.6(a). On the other hand, the right-hand side of Eq. (6.12) presents the kinetic energy at the zero-restoring-force point. From Eq. (6.12), v_c can be expressed in terms of u_p by

$$v_c/V_y = \sqrt{(\alpha/4)(u_p/d_y)^2 + (2\alpha - 1)(u_p/d_y) + (1/\alpha)} \qquad (6.13)$$

The plastic deformation amplitude u_p after each impulse can be obtained from the energy balance law. The energy balance law between the zero-restoring-force point (point A in Figure 6.6) and the point attaining the maximum deformation (point B in Figure 6.6) can be derived as

$$m(v_c + V)^2/2 = \alpha k\{(f_y + 0.5\alpha k u_p)/(\alpha k)\}^2/2 \tag{6.14}$$

The left-hand side of Eq. (6.14) expresses the kinetic energy computed by the velocity $(v_c + V)$ of mass just after each impulse. On the other hand, the right-hand side of Eq. (6.14) presents the hysteretic and elastic strain energy indicated by the shaded area including points A, B in Figure 6.6(a).

Substitution of Eq. (6.13) into Eq. (6.14) and rearrangement of the resulting equation provide

$$u_p/d_y = \{(V/V_y)^2 - 2(V/V_y)/\sqrt{\alpha}\}/\{2\alpha - 2 + \sqrt{\alpha}(V/V_y)\} \tag{6.15}$$

From Eq. (6.15) and Figure 6.6(a), u_{max} can be obtained as follows:

$$u_{max}/d_y = 1 + 0.5(u_p/d_y)$$
$$= 1 + 0.5\left[\{(V/V_y)^2 - 2(V/V_y)/\sqrt{\alpha}\}/\{2\alpha - 2 + \sqrt{\alpha}(V/V_y)\}\right] \tag{6.16}$$

From Eq. (6.15) or Eq. (6.16), the elastic-plastic response diverges to infinity under the condition of $2\alpha - 2 + \sqrt{\alpha}(V/V_y) = 0$. In CASE 2, the impulse input velocity level at which the response diverges can be obtained from $2\alpha - 2 + \sqrt{\alpha}(V/V_y) = 0$ as follows:

$$V/V_y = (-2\alpha + 2)/\sqrt{\alpha} \tag{6.17}$$

The response divergence phenomenon can occur under the condition $V/V_y \geq (-2\alpha + 2)/\sqrt{\alpha}$ because the increment of the input energy due to the repetitive impulses cannot be consumed by plastic deformation. The same phenomenon was observed before under a sinusoidal wave input (Iwan 1961).

From Eqs. (6.11) and (6.17), the input velocity level in CASE 2 has to satisfy the following inequality condition.

$$-2 + 2\sqrt{1/\alpha} < V/V_y < (-2\alpha + 2)/\sqrt{\alpha} \tag{6.18}$$

6.3.3 Results in numerical example

The plastic deformation amplitudes u_p/d_y obtained in Sections 6.3.1 and 6.3.2 are shown in Figures 6.7(a), (b). Figure 6.7(a) illustrates the normalized plastic deformation amplitude u_p/d_y with respect to the input velocity level V/V_y for various post-yield stiffness ratios $\alpha = 0, 0.1, 0.2, 0.3, 0.4, 0.5, 0.6, 0.7, 0.8, 0.9$. On the other hand, Figure 6.7(b) shows the plastic deformation amplitude u_p/d_y with respect to the post-yield stiffness ratio α for

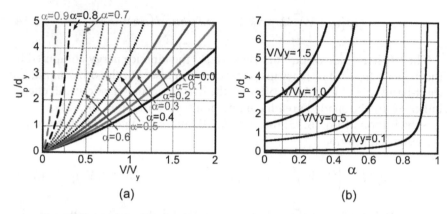

Figure 6.7 Plastic deformation amplitude u_p/d_y to critical multi-impulse: (a) u_p/d_y with respect to input level V/V_y for various post-yield stiffness ratios α, (b) u_p/d_y with respect to post-yield stiffness ratio α for various input levels V/V_y. (Kojima and Takewaki 2017).

various input velocity levels $V/V_y = 0.1, 0.5, 1.0, 1.5$. The model with $\alpha = 0$ is equivalent to the EPP SDOF model and u_p/d_y for this model was derived in Kojima and Takewaki (2015c) and Chapter 4.

6.3.4 Derivation of critical impulse timing

The time intervals between two consecutive impulses are derived in this section for CASE 1 and CASE 2. In CASE 1 and CASE 2, each impulse acts at the zero-restoring-force point. It is interesting to note that when the multi-impulses are acted sequentially, the time intervals are not constant in the first several cycles. Then the time intervals converge to a constant value. The time interval t_0^c between two consecutive impulses can be obtained by solving the equations of motion and substituting the continuation conditions at the transition points in the steady state. The time interval t_0^c, shown in Figure 6.4(b) and Figure 6.6(b), can be expressed as follows:

$$
\begin{aligned}
\frac{t_0^c}{T_1} = \frac{1}{2\pi} & \left[\arcsin\left\{ \frac{1 - 0.5\alpha\left(u_p/d_y\right)}{(v_c + V)/V_y} \right\} \right. \\
& \left. + \frac{1}{\sqrt{\alpha}} \arctan\left\{ \frac{1}{\sqrt{\alpha}} \frac{v_B/V_y}{(1/\alpha) - 0.5\left(u_p/d_y\right)} \right\} \right] + \frac{1}{4} \\
& \text{for } V/V_y \le -2 + 2\sqrt{1/\alpha}
\end{aligned}
\tag{6.19a}
$$

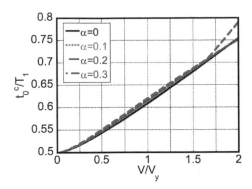

Figure 6.8 Critical impulse timing t_0^c/T_1 with respect to input level V/V_y for various post-yield stiffness ratios α (Kojima and Takewaki 2017).

$$\frac{t_0^c}{T_1} = \frac{1}{4}\left(1+\frac{1}{\sqrt{\alpha}}\right)+\frac{1}{2\pi}\left[-\arcsin\left\{\frac{0.5\alpha\left(u_p/d_y\right)-1}{0.5\alpha\left(u_p/d_y\right)+1}\right\}\right.$$
$$\left.+\frac{1}{\sqrt{\alpha}}\arctan\left\{\frac{0.5\left(u_p/d_y\right)-\left(1/\alpha\right)}{\sqrt{2\left(u_p/d_y\right)}}\right\}\right] \qquad (6.19b)$$

for $V/V_y > -2+2\sqrt{1/\alpha}$

The normalized quantities u_p/d_y and v_c/V_y in Eq. (6.19a) can be obtained from Eqs. (6.4) and (6.2), and u_p/d_y in Eq. (6.19b) can be obtained from Eq. (6.15). In addition, the velocity v_B/V_y at point B in Eq. (6.19a) is obtained by

$$v_B/V_y = \sqrt{\left\{\left(v_c+V\right)/V_y\right\}^2-\left\{1-0.5\alpha\left(u_p/d_y\right)\right\}^2} \quad \text{for } V/V_y \le -2+2\sqrt{1/\alpha} \quad (6.20)$$

The detailed derivation of Eqs. (6.19a), (6.19b), and (6.20) is shown in Appendix 1.

Figure 6.8 presents the normalized quantity of the time interval t_0^c with respect to the input velocity level for various post-yield stiffness ratios $\alpha = 0$, 0.1, 0.2, 0.3. The model with $\alpha = 0$ is equivalent to the EPP SDOF model and t_0^c in this model was derived in Kojima and Takewaki (2015c) and Chapter 4.

6.4 CONVERGENCE OF CRITICAL IMPULSE TIMING

In this section, it is investigated whether the response under the multi-impulse with the equal time interval t_0^c obtained in Section 6.3.4 converges to the steady state in which each impulse acts at the zero restoring-force point as shown in Figure 6.3. Another possibility is to act the multi-impulses at the zero-restoring-force timing. In this case, the time intervals of the multi-impulses are not constant at the beginning and may converge to a constant. In this section, the former analysis is made.

The closed-form expression for the time history response in the steady state can be derived (see Appendix 1). However, the transient response is complicated because the number of impulses for assuring convergence depends on the input velocity level and the post-yield stiffness ratio. The time-history response analysis is used to calculate the response under the multi-impulse with the time interval t_0^c . The parameters $T_1 = 1.0$ [sec], $d_y = 0.04$ [m], $\Delta t = 1.0 \times 10^{-4} T_1$ were used in the analysis. The parameter Δt denotes the time increment used in the time-history response analysis. The response under the multi-impulse is calculated by adding $\pm V$ to the velocity of the mass at the impulse acting timing. Figures 6.9, 6.10, and 6.11 show

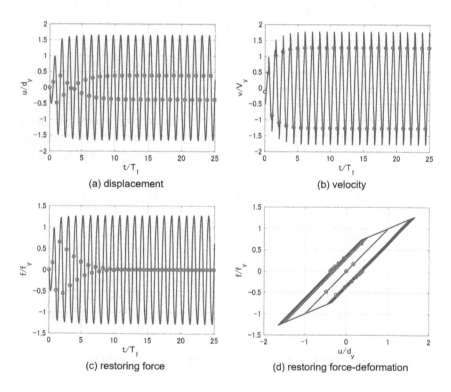

Figure 6.9 Response under multi-impulse with time interval t_0^c for $V/V_y = 0.5$ and $\alpha = \tan(\pi/8) = 0.414$ (impulse timing is critical one): (a) displacement, (b) velocity, (c) restoring force, and (d) restoring force-deformation relation (circles indicate acting points of impulses) (Kojima and Takewaki 2017).

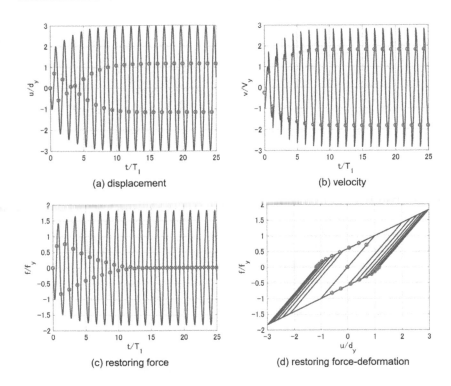

Figure 6.10 Response under multi-impulse with time interval t_0^c for $V/V_y = 1.0$ and $\alpha = \tan(\pi/8) = 0.414$ (impulse timing is critical one): (a) displacement, (b) velocity, (c) restoring force, and (d) restoring force-deformation relation (circles indicate acting points of impulses) (Kojima and Takewaki 2017).

the time histories of relative displacement, relative velocity, restoring force, and restoring force-deformation relation under the multi-impulse with the time interval t_0^c in the model with $\alpha = \tan(\pi/8) = 0.414$ for $V/V_y = 0.5, 1.0,$ 1.5. This post-yield stiffness ratio was taken from the comparative past work (Iwan 1961). The time interval used in this section was obtained by using the assumption of the steady state. The circles in Figures 6.9, 6.10, and 6.11 indicate the acting points of impulses. It can be seen that the response converges to a state in which each impulse acts at the zero restoring force irrespective of the input velocity level, and the maximum deformation and the plastic deformation amplitude after convergence correspond to the closed-form expressions obtained in Section 6.3.1 and Section 6.3.2. In the model with $\alpha = \tan(\pi/8) = 0.414$, the input velocity levels $V/V_y = 0.5, 1.0$ correspond to CASE 1 in Section 6.3.1, and the acting points of impulses converge to the zero restoring force timing in the unloading process in Figures 6.9 and 6.10. From Figures 6.9 and 6.10, the required number of impulses for assuring convergence is about 25. On the other hand, the input

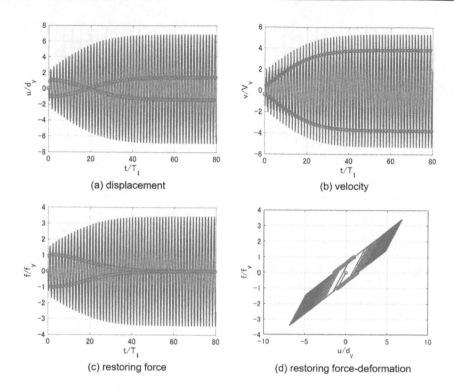

Figure 6.11 Response under multi-impulse with time interval t_0^c for $V/V_y = 1.5$ and $\alpha = \tan(\pi/8) = 0.414$ (impulse timing is critical one): (a) displacement, (b) velocity, (c) restoring force, and (d) restoring force-deformation relation (circles indicate acting points of impulses) (Kojima and Takewaki 2017).

velocity level $V/V_y = 1.5$ corresponds to CASE 2 in Section 6.3.2 and the acting points of impulses converge to the zero-restoring-force timing in the loading process in Figure 6.11. From Figure 6.11, CASE 2 requires over 100 impulses for convergence.

6.5 ACCURACY CHECK BY TIME-HISTORY RESPONSE ANALYSIS TO CORRESPONDING MULTI-CYCLE SINUSOIDAL WAVE

In order to check the accuracy of using the multi-impulse with the equal time interval as a substitute of the corresponding multi-cycle sinusoidal wave representing long-duration ground motions, the time-history response analysis of the bilinear hysteretic SDOF system under the corresponding multi-cycle sinusoidal wave is conducted.

In the present evaluation, it is important to adjust the input level of the multi-impulse and the corresponding multi-cycle sinusoidal wave based on the equivalence of the maximum Fourier amplitude (see Appendix 2 in this chapter and Appendix 1 in Chapter 4). It is noted that although two different multi-impulses with modification in the first impulse are treated in Chapter 4 and in this chapter, the resulting relation of the multi-impulse and the multi-cycle sinusoidal waves is the same. The period, the circular frequency, the acceleration amplitude, and the velocity amplitude of the corresponding sinusoidal wave are denoted by T_l, $\omega_l = 2\pi/T_l$, A_l, and $V_l = A_l/\omega_l$, respectively, and $T_l = 2t_0{}^c$ is used in this section. The number of cycles of the multi-cycle sinusoidal wave is a half of the number of impulses. In the derivation of the response under the multi-impulse, the steady state after a sufficient number of impulses is assumed as shown in Figures 6.9, 6.10, and 6.11. The relation between the input velocity level of the multi-impulse with the sufficient number of impulses (for example, over 20 impulses) and the acceleration amplitude of the corresponding multi-cycle sinusoidal wave with the sufficient number of cycles is expressed as follows (see Appendix 2 in this chapter and Appendix 1 in Chapter 4):

$$V_l = A_l/\omega_l = (2/\pi)V \tag{6.21}$$

The derivation of Eq. (6.21) was shown in Appendix 1 in Chapter 4 and in Appendix 2 of this chapter.

Figure 6.12 presents the comparison of the plastic deformation amplitude and the maximum deformation normalized by the yield deformation of the bilinear hysteretic SDOF system under the multi-impulse and the corresponding multi-cycle sinusoidal wave with respect to input velocity level. The response under the multi-impulse comes from the closed-form expressions derived in Sections 6.3.1 and 6.3.2, and the response under the corresponding multi-cycle sinusoidal wave is obtained by using the time-history response analysis. The structural and computational parameters $T_1 = 1.0$ [sec], $d_y = 0.04$ [m], $\Delta t = 1.0 \times 10^{-4}T_1$ are used in the time-history response analysis, and the numbers of cycles used in the time-history response analysis are 100 cycles for $\alpha = \tan(2\pi/180) = 0.035$, 500 cycles for $\alpha = \tan(\pi/8) = 0.414$, and 1000 cycles for $\alpha = 0.9$. These post-yield stiffness ratios were taken from the past comparable work (Iwan 1961). The multi-impulse provides a fairly good result for the multi-cycle sinusoidal wave in the evaluation of the maximum deformation and the plastic deformation amplitude if the maximum Fourier amplitude is adjusted. In order to relate the elastic-plastic responses under the multi-cycle sinusoidal wave to that under the multi-impulse, it is necessary to amplify the acceleration amplitude of the corresponding multi-cycle sinusoidal wave by 1.15 after both Fourier amplitudes of the sinusoidal wave and the multi-impulse are adjusted in the model with the elastic–perfectly plastic restoring force characteristics ($\alpha = 0$) (Kojima and Takewaki 2015c and

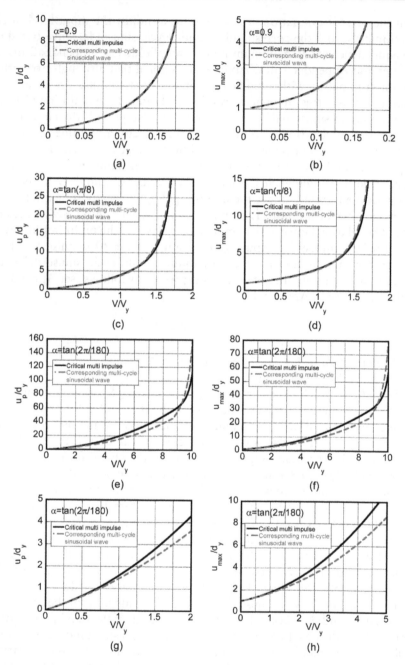

Figure 6.12 Comparison of plastic deformation and maximum deformation between critical multi impulse and corresponding multi-cycle sinusoidal wave: (a, b) α = 0.9, (c, d) α = tan $(\pi/8)$ = 0.414, (e, f, g, h) α = tan $(2\pi/180)$ = 0.0349 ((g) and (h) are magnified ones of (e) and (f)) (Kojima and Takewaki 2017).

Chapter 4 in this book). The maximum deformation under the multi-impulse is larger than that under the corresponding multi-cycle sinusoidal wave in $V/V_y < -2 + \sqrt{1/\alpha}$ in CASE 1. On the other hand, the maximum deformation under the corresponding multi-cycle sinusoidal wave is larger than that under the multi-impulse in $V/V_y > -2 + \sqrt{1/\alpha}$ in CASE 2.

6.6 PROOF OF CRITICAL TIMING

To investigate the validity of the critical timing evaluated by Eq. (6.19a, b), the time-history response analysis was conducted for the bilinear hysteretic SDOF system under the multi-impulse with the varied impulse timing t_0 for various input velocity levels and various post-yield stiffness ratios. The critical timing of each impulse can be characterized as the time with zero restoring force as assumed in Section 6.3. The structural and computational parameters $T_1 = 1.0$ [sec], $d_y = 0.04$ [m], $\Delta t = 1.0 \times 10^{-4} T_1$ were used in the time-history response analysis, and the numbers of impulses used in the time-history response analysis for the convergence of the response are 1,000.

Figure 6.13 shows the normalized maximum deformation u_{max}/d_y and the normalized plastic deformation amplitude u_p/d_y with respect to the impulse timing $t_0/t_0{}^c$ normalized by the critical timing for various input velocity levels V/V_y and various post-yield stiffness ratios $\alpha = 0.035, 0.414, 0.9$. It can be found that the critical timing $t_0{}^c$ derived in Section 6.3.4 actually provides the critical case under the multi-impulse and gives the upper bound of u_{max}/d_y and u_p/d_y. The closed-form expressions for u_{max}/d_y and u_p/d_y derived in Sections 6.3.1 and 6.3.2 are equal to the upper bound of u_{max}/d_y and u_p/d_y in Figure 6.13.

6.7 APPLICABILITY OF CRITICAL MULTI-IMPULSE TIMING TO CORRESPONDING SINUSOIDAL WAVE

In Section 6.5, it was demonstrated that if the maximum value of the Fourier amplitude is selected as a key parameter, the response under the multi-impulse with the time interval obtained by Eq. (6.19a, b) and that under the corresponding multi-cycle sinusoidal wave exhibit a good correspondence. In this section, it is investigated whether the critical timing of the multi-impulse derived in Section 6.3.4 is also appropriate in comparison with the result for the multi-cycle sinusoidal wave.

The resonant equivalent frequency of the harmonic wave for a specific acceleration amplitude has to be evaluated by the resonance curve computed by using the exact solution (Iwan 1961). In this procedure, it is necessary to solve the transcendental equation by changing the excitation frequency parametrically. On the other hand, Caughey (1960) has proposed the method to derive the equivalent resonance frequency directly by using

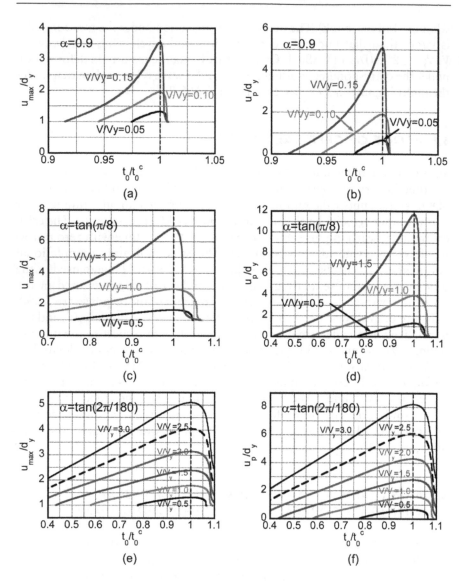

Figure 6.13 Maximum deformation and plastic deformation amplitude with respect to timing of multi-impulse for various input levels: (a, b) $\alpha = 0.9$, (c, d) $\alpha = \tan(\pi/8) = 0.414$, (e, f) $\alpha = \tan(2\pi/180) = 0.0349$ (Kojima and Takewaki 2017).

the equivalent linearization method with the least-squares approximation. However, this equivalent resonant frequency differs from the exact equivalent resonant frequency in the larger acceleration amplitude range. In these previous papers, the resonant equivalent frequency of the harmonic wave for a specific "acceleration" amplitude was derived. However, the resonant equivalent frequency for a specific "velocity" amplitude was not derived.

Recently, it has been recommended to normalize earthquake ground motions with the velocity amplitude (Takewaki and Tsujimoto 2011).

In order to calculate the maximum deformation and the plastic deformation amplitude under the corresponding multi-cycle sinusoidal wave with the varied period T_l for various input velocity levels and various post-yield stiffness ratios, the time-history response analysis was conducted for the bilinear hysteretic SDOF system under the corresponding multi-cycle sinusoidal wave. The parameters T_l, $\omega_l = 2\pi/T_l$, A_l, and $V_l = A_l/\omega_l$ denote the period, the circular frequency, the acceleration amplitude, and the velocity amplitude of the sinusoidal wave corresponding to the multi-impulse with the equal time interval t_0 and the input velocity level V, respectively. In addition, $T_l = 2t_0$ is used in this section. The input period T_l is changed for the specific velocity amplitude calculated by Eq. (6.21) with the input velocity level V. $T_l^c = 2t_0^c$ denotes the approximate critical period of the multi-cycle sinusoidal wave for a specific velocity amplitude V_l.

Figure 6.14 shows the normalized maximum deformation u_{max}/d_y and the normalized plastic deformation amplitude u_p/d_y with respect to the input period $T_l/T_l^c(=t_0/t_0^c)$ normalized by the approximate critical period for various input velocity levels V/V_y (corresponding to the velocity amplitude V_l) and various post-yield stiffness ratios $\alpha = 0.035, 0.414, 0.9$. It can be seen that $T_l^c = 2t_0^c$ is a good approximation of the critical period of the multi-cycle sinusoidal wave for a specific velocity amplitude.

6.8 ACCURACY CHECK BY EXACT SOLUTION TO CORRESPONDING MULTI-CYCLE SINUSOIDAL WAVE

The accuracy of the present closed-form steady-state response under the critical multi-impulse is investigated in this section through the comparison with the resonance curve under the corresponding sinusoidal wave computed by using the exact solution (Iwan 1961). For drawing the resonance curve, it is necessary to solve the transcendental equation by changing the excitation frequency parametrically, and the resonant equivalent frequency of the harmonic wave for a specific acceleration amplitude has to be evaluated by the resonance curve (Iwan 1961). On the other hand, the method explained in this chapter provides the critical steady-state response for the specific input level by the closed-form expression directly. The input level of the multi-impulse and the corresponding sinusoidal wave was adjusted by using the equivalence of the maximum Fourier amplitude as explained in Sections 6.5 and 6.7 (Appendix 1 in Chapter 4).

Figure 6.15 shows the comparison of the present closed-form expression for the critical maximum deformation with respect to ω^* with the resonance curve by Iwan (1961) (corresponding to figures 11–13 in Iwan (1961)). The parameters ω^* and r in Figure 6.15 denote the ratio of the excitation

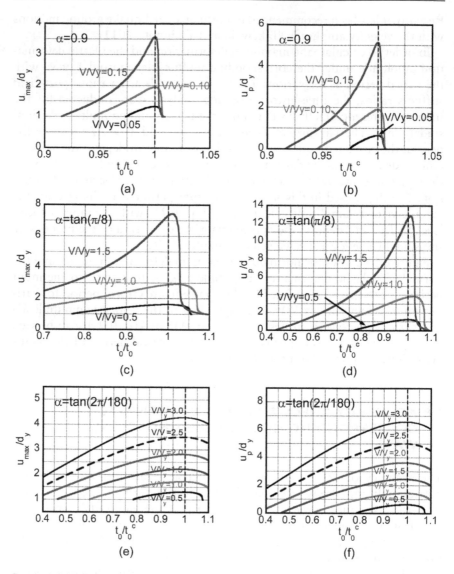

Figure 6.14 Maximum deformation and plastic deformation amplitude with respect to period of corresponding sinusoidal wave for various input levels: (a, b) $\alpha = 0.9$, (c, d) $\alpha = \tan(\pi/8) = 0.414$, (e, f) $\alpha = \tan(2\pi/180) = 0.0349$ (Kojima and Takewaki 2017).

frequency $\omega_l = 2\pi/T_l$ of the corresponding sinusoidal wave to the elastic natural circular frequency ω_1 and the ratio of the excitation acceleration amplitude $A_l = \omega_l V_l$ of the corresponding sinusoidal wave to the parameter $A_y = \omega_1^2 d_y$. The parameter r is also equal to the product of the mass m and the acceleration amplitude A_l normalized by the yield force f_y. The backbone curve in Figure 6.15 shows the maximum deformation under the

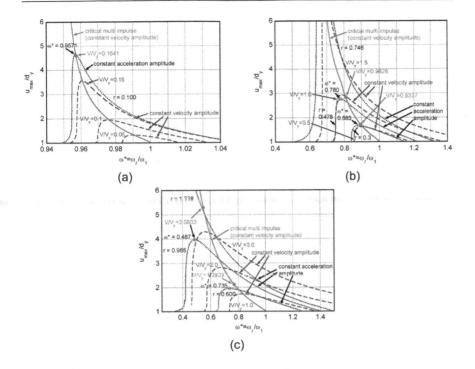

Figure 6.15 Comparison of closed-form maximum deformation to critical multi-impulse (constant velocity amplitude) and resonance curve to sinusoidal wave (constant acceleration amplitude and constant velocity amplitude): (a) $\alpha = 0.9$, (b) $\alpha = \tan(\pi/8) = 0.414$, (c) $\alpha = \tan(2\pi/180) = 0.0349$ (Kojima and Takewaki 2017).

critical multi-impulse. The normalized critical timing $t_0{}^c/T_1$ is converted to $\omega^* = T_1/(2t_0{}^c)$ by using $T_l = 2t_0{}^c$ in the critical case. The dotted line (with peaks) shows the resonance curve with $r = 0.1$ in Figure 6.15(a), $r = 0.3$, 0.478, 0.746 in Figure 6.15(b), and $r = 0.6$, 0.955, 1.228 in Figure 6.15(c). The open circles on the dotted lines in Figure 6.15 present the resonance points for the specific acceleration amplitude. In addition, the broken line in Figure 6.15 presents the resonance curve for constant velocity amplitude. It can be observed that the present closed-form expression for the critical maximum deformation under the multi-impulse corresponds to the broken line (constant velocity amplitude) better than the dotted line (constant acceleration amplitude).

The solid circles present the maximum deformation under the critical multi-impulse for the input levels corresponding to the resonance points of the resonance curve (the open circles in Figure 6.15). The method to calculate the input velocity level corresponding to the resonant point (open circle) is explained in the following.

Table 6.1 Comparison of maximum deformations to sinusoidal wave and multi-impulse (Kojima and Takewaki 2017)

α	Resonant response subjected to sinusoidal wave (exact solution by Iwan 1961)			Close-form solution subjected to critical multiple impulse (corresponding to resonant point of exact solution)			
	$r = A_l/A_y$	$\omega = \omega_l/\omega_1$	u_{max}/d_y	V/V_y: Eq. (6.23)	t_0^c/T_1: Eq. (6.19a) or Eq.(6.19b)	$\omega^* = T_1/(2t_0^c)$	u_{max}/d_y: Eq. (6.5) or Eq.(6.16)
0.9	0.100	0.957	4.645	0.1641	0.5223	0.9573	4.603
$\tan(\pi/8)$	0.478	0.780	2.756	0.9626	0.6191	0.8077	2.845
= 0.414	0.300	0.883	1.676	0.5337	0.5590	0.8945	1.711
$\tan(2°)$	0.955	0.487	3.972	3.0803	0.9115	0.5486	5.293
= 0.035	0.600	0.735	1.952	1.2823	0.6500	0.7693	2.116

From given parameters r, ω^* (at the resonance point) and Eq. (6.21), the following relation can be obtained.

$$V_l = A_l/\omega_l = (r\omega_1 V_y)/(\omega^* \omega_1) = (2/\pi)V \qquad (6.22)$$

From Eq. (6.22), the normalized input velocity level can be obtained as follows:

$$V/V_y = (\pi/2)(r/\omega^*) \qquad (6.23)$$

The maximum deformation u_{max}/d_y and the critical timing t_0^c/T_1 can be obtained from Eq. (6.5) or Eq. (6.16) and Eq. (6.19a) or Eq. (6.19b) depending on the input velocity level, respectively.

The results of the correspondence between the critical multi-impulse and the critical sinusoidal wave are listed in Table 6.1. The responses and the resonant frequencies obtained for the critical multi-impulse and the critical sinusoidal wave exhibit fairly good correspondence except for the case with $\alpha = \tan(2\pi/180)$ and $r = 0.600$ (the small post-yield stiffness with the large input level).

6.9 SUMMARIES

The multi-impulse was introduced as a substitute for the long-duration ground motion as in Chapter 4, and the closed-form expression was derived for the steady-state elastic-plastic response of an undamped bilinear hysteretic SDOF system under the critical multi-impulse. While the computation of the resonant equivalent frequency of the elastic-plastic system is a tough task in the conventional method dealing directly with the sinusoidal wave (Caughey 1960a, b, Iwan 1961), the steady-state elastic-plastic response under the critical multi-impulse can be obtained explicitly (without repetition) and the critical time interval of the multi-impulse (the resonant frequency) can also be obtained explicitly for the increasing input level. The conclusions can be summarized as follows.

1. The existence of the steady state in which each impulse acts at the zero-restoring-force point was assumed, and the closed-form expressions were derived for the elastic-plastic response under the critical multi-impulse by using the energy balance approach. The steady state under the critical multi-impulse can be classified into two cases depending on the plastic deformation and the input velocity level. CASE 1 is the case where each impulse acts at the zero-restoring-force timing in the unloading process. On the other hand, CASE 2 is the case where each impulse acts at the zero-restoring-force timing in the loading process. The closed-form expressions for the critical time interval of the multi-impulse in both CASE 1 and CASE 2 were derived by solving the equations of motion and substituting the continuation conditions at the transition points.

2. The response under the multi-impulse with the critical equal time interval obtained in Section 6.3.4 converges to the steady state in which each impulse acts at the zero restoring force as shown in Figure 6.3. The maximum deformation and the plastic deformation amplitude after convergence to the steady state correspond to the closed-form expressions obtained in Sections 6.3.1 and 6.3.2.

3. The validity and accuracy of the obtained closed-form expressions were investigated through the comparison with the steady-state response under the corresponding multi-cycle sinusoidal wave as a representative of the long-duration ground motion by using the time-history response analysis. It was confirmed that the multi-impulse provides a fairly good substitute of the multi-cycle sinusoidal wave in the evaluation of the maximum deformation and the plastic deformation amplitude if the maximum Fourier amplitude is adjusted in a reasonable manner.

4. The validity of the critical time interval derived in Section 6.3.4 was confirmed by using the time-history response analysis of the bilinear hysteretic SDOF system under the multi-impulse with the varied impulse timing. The critical timing of each impulse can be characterized by the time with zero restoring force in the steady state.

5. Twice the critical time interval is a good approximate of the critical period of the multi-cycle sinusoidal wave with the corresponding input amplitude for attaining the maximum response.

In this chapter, the closed-form expression for the critical elastic-plastic response was derived for a specific input velocity level V of the multi-impulse. The input velocity level V corresponds to the velocity amplitude of the long-duration ground motion. The earthquake ground motions have been recorded for 70–80 years all over the world, and the most rational method in determining V is to predict the velocity amplitude and the period of the ground motion at a specific site from the magnitude and/or other parameters of the possible fault rupture. However, it seems quite difficult to predict a possible ground motion at a specific site even by the most advanced method (Makita et al. 2018a, b, 2019, Kondo and Takewaki 2019). In such a situation, the most reliable method may be to determine the input velocity level V from the occurrence return period of ground motions and the level of importance of the object building structure. In this case, an allowable level of damage to the structure should be set depending on the level of importance of the structure. From an alternative viewpoint, the following treatment may be possible. The relation between V/V_y and the ductility factor u_{max}/d_y was obtained as a result in this chapter. If two of V, V_y, and u_{max}/d_y are given, the remaining one can be obtained. V_y represents the strength and stiffness parameter of the structure in velocity dimension. Therefore, if two among the structural parameter V_y (the strength and stiffness parameter of the structure), the input level V of the ground motion, and the allowable damage level u_{max}/d_y of the structure are given, the remaining parameter can be determined. The final decision is entrusted to structural designers. It may be said that the method explained in this chapter offered a tool for such a decision.

REFERENCES

Abrahamson, N., Ashford, S., Elgamal, A., Kramer, S., Seible, F., and Somerville, P. (1998). *1st PEER Workshop on Characterization of Special Source Effects*, Pacific Earthquake Engineering Research Center, University of California, San Diego.

Bertero, V. V., Mahin, S. A., and Herrera, R. A. (1978). Aseismic design implications of near-fault San Fernando earthquake records, *Earthquake Eng. Struct. Dyn.*, 6(1), 31–42.

Caughey, T. K. (1960a). Sinusoidal excitation of a system with bilinear hysteresis, *J. Appl. Mech.*, 27, 640–643.

Caughey, T. K. (1960b). Random excitation of a system with bilinear hysteresis, *J. Appl. Mech.*, 27, 649–652.

Iwan, W. D. (1961). The dynamic response of bilinear hysteretic systems, Ph.D. Thesis, California Institute of Technology, Pasadena.

Iwan, W. D. (1965a). The dynamic response of the one-degree-of-freedom bilinear hysteretic system, *Proc. of the Third World Conf. on Earthquake Eng.*, New Zealand.

Iwan, W. D. (1965b). The steady-state response of a two-degree-of-freedom bilinear hysteretic system, *J. Applied Mech.*, 32(1), 151–156.

Kalkan, E., and Kunnath, S. K. (2006). Effects of fling step and forward directivity on seismic response of buildings, *Earthquake Spectra*, 22(2), 367–390.

Kojima, K., and Takewaki, I. (2015a). Critical earthquake response of elastic-plastic structures under near-fault ground motions (Part 1: Fling-step input), *Frontiers in Built Environment*, 1: 12.

Kojima, K., and Takewaki, I. (2015b). Critical earthquake response of elastic-plastic structures under near-fault ground motions (Part 2: Forward-directivity input), *Frontiers in Built Environment*, 1: 13.

Kojima, K., and Takewaki, I. (2015c). Critical input and response of elastic-plastic structures under long-duration earthquake ground motions, *Frontiers in Built Environment*, 1: 15.

Kojima, K., and Takewaki, I. (2016). Closed-form critical earthquake response of elastic-plastic structures with bilinear hysteresis under near-fault ground motions, *J. Struct. Construction Eng.*, AIJ, 726, 1209–1219 (in Japanese).

Kojima, K., and Takewaki, I. (2017). Critical steady-state response of SDOF bilinear hysteretic system under multi impulse as substitute of long-duration ground motions, *Frontiers in Built Environment*, 3: 41.

Kondo, K., and Takewaki, I. (2019). Simultaneous approach to critical fault rupture slip distribution and optimal damper placement for resilient building design, *Frontiers in Built Environment*, 5: 126.

Liu, C.-S. (2000). The steady loops of SDOF perfectly elastoplastic structures under sinusoidal loadings, *J. Mar. Sci. Technol.*, 8, 50–60.

Makita, K., Murase, M., Kondo, K., and Takewaki, I. (2018a). Robustness evaluation of base-isolation building-connection hybrid controlled building structures considering uncertainties in deep ground, *Frontiers in Built Environment*, 4: 16.

Makita, K., Kondo, K., and Takewaki, I. (2018b). Critical ground motion for resilient building design considering uncertainty of fault rupture slip, *Frontiers in Built Environment*, 4: 64.

Makita, K., Kondo, K., and Takewaki, I. (2019). Finite difference method-based critical ground motion and robustness evaluation for long-period building structures under uncertainty in fault rupture, *Frontiers in Built Environment*, 5: 2.

Roberts, J. B., and Spanos, P. D. (1990). *Random Vibration and Statistical Linearization*. New York: Wiley.

Takewaki, I., Murakami, S., Fujita, K., Yoshitomi, S. and Tsuji, M. (2011). The 2011 off the Pacific coast of Tohoku earthquake and response of high-rise buildings under long-period ground motions, *Soil Dyn. Earthquake Eng.*, 31(11), 1511–1528.

Takewaki, I., and Tsujimoto, H. (2011). Scaling of design earthquake ground motions for tall buildings based on drift and input energy demands, *Earthquakes Struct.*, 2(2), 171-187.

Takewaki, I., Fujita, K., and Yoshitomi, S. (2013). Uncertainties in long-period ground motion and its impact on building structural design: Case study of the 2011 Tohoku (Japan) earthquake, *Engineering Structures*, 49, 119–134.

APPENDIX 1: TIME-HISTORY RESPONSE TO CRITICAL MULTI-IMPULSE AND DERIVATION OF CRITICAL TIME INTERVAL

The closed-form expressions for the time-history response under the critical multi-impulse and the critical time interval in the steady state are derived here by solving the equation of motion directly.

First of all, the time-history response for CASE 1 is derived. Figures 6.4(b), (c) show the time histories of the deformation and the restoring force in CASE 1. By solving the equation of motion in the path between point F and B in Figure 6.4(a) and substituting the displacement and velocity conditions at point A, the time-history response after the impulse acting point (point A in Figure 6.4(a)) can be expressed by:

$$u(t) = \{(v_c + V)/V_y\} d_y \sin(\omega_1 t) - 0.5(1 - \alpha) u_p \qquad (6.\text{A1a})$$

$$\dot{u}(t) = (v_c + V)\cos(\omega_1 t) \qquad (6.\text{A1b})$$

In Eqs. (6.A1a) and (6.A1b), $t = 0$ is set at point A and v_c, u_p can be obtained from Eqs. (6.2) and (6.4). The time interval between points A and B in Figure 6.4 is denoted by t_{AB} as shown in Figures 6.4(b), (c). t_{AB} can then be obtained as follows from $u(t = t_{AB}) = d_y - 0.5 u_p$ and Eq. (6.A1a).

$$t_{AB}/T_1 = \{1/(2\pi)\} \arcsin\left[\{1 - 0.5\alpha (u_p/d_y)\}/\{(v_c + V)/V_y\}\right] \qquad (6.\text{A2})$$

The time-history response after the yielding point (point B in Figure 6.4) can be expressed as follows:

$$u(t) = \left(\frac{1}{\alpha} - 0.5\frac{u_p}{d_y}\right) d_y \cos\left(\sqrt{\alpha}\,\omega_1 t\right)$$
$$+ \frac{1}{\sqrt{\alpha}}\frac{v_B}{V_y} d_y \sin\left(\sqrt{\alpha}\,\omega_1 t\right) - \left(\frac{1}{\alpha} - 1\right) d_y \qquad (6.\text{A3a})$$

$$\dot{u}(t) = -\sqrt{\alpha\left(\frac{1}{\alpha} - 0.5\frac{u_p}{d_y}\right)^2 + \left(\frac{v_B}{V_y}\right)^2}$$
$$\times V_y \sin\left[\sqrt{\alpha}\,\omega_1 t - \arctan\frac{v_B/V_y}{\sqrt{\alpha}\{(1/\alpha) - 0.5(u_p/d_y)\}}\right] \qquad (6.\text{A3b})$$

In Eqs. (6.A3a) and (6.A3b), $t = 0$ is set at point B. The velocity v_B at point B can then be obtained as shown in Eq. (6.20) by the following energy balance law between point A and point B.

$$m\left(v_c + V\right)^2 / 2 = \left(m v_B^2 / 2\right) + \left\{\left(f_y - 0.5\alpha k u_p\right)^2 / 2k\right\} \tag{6.A4}$$

The time interval between points B and C in Figure 6.4 is denoted by t_{BC} as shown in Figures 6.4(b), (c). t_{BC} can then be obtained as follows from $\dot{u}(t = t_{BC}) = 0$ and Eq. (6.A3b).

$$t_{BC}/T_1 = \left\{1/\left(2\pi\sqrt{\alpha}\right)\right\} \arctan\left[\frac{v_B/V_y}{\sqrt{\alpha}\left\{(1/\alpha) - 0.5\left(u_p/d_y\right)\right\}}\right] \tag{6.A5}$$

The time-history response after the unloading starting point (point C in Figure 6.4) can be expressed as follows:

$$u(t) = \left\{1 + 0.5\alpha\left(u_p/d_y\right)\right\} d_y \cos(\omega_1 t) + 0.5(1 - \alpha) u_p \tag{6.A6a}$$

$$\dot{u}(t) = -\left\{1 + 0.5\alpha\left(u_p/d_y\right)\right\} V_y \sin(\omega_1 t) \tag{6.A6b}$$

In Eqs. (6.A6a) and (6.A6b), $t = 0$ is set at point C. The time interval between points C and D in Figure 6.4 is denoted by t_{CD} as shown in Figures 6.4(b), (c). t_{CD} can then be obtained as follows from $u(t = t_{CD}) = 0.5(1 - \alpha)u_p$ and Eq. (6.A6a).

$$t_{CD}/T_1 = 0.25 \tag{6.A7}$$

From Eqs. (6.A2), (6.A5), and (6.A7) and Figures 6.4(b), (c), the time interval t_0^c between two consecutive impulses acting at the zero-restoring-force points (points A and D) in CASE 1 can be obtained as follows:

$$\begin{aligned}
t_0^c/T_1 &= \left(t_{AB}/T_1\right) + \left(t_{BC}/T_1\right) + \left(t_{CD}/T_1\right) \\
&= \frac{1}{2\pi}\left[\arcsin\left\{\frac{1 - 0.5\alpha\left(u_p/d_y\right)}{\left(v_c + V\right)/V_y}\right\}\right. \\
&\quad \left. + \frac{1}{\sqrt{\alpha}}\arctan\left\{\frac{1}{\sqrt{\alpha}}\frac{v_B/V_y}{(1/\alpha) - 0.5\left(u_p/d_y\right)}\right\}\right] + \frac{1}{4} \tag{6.A8} \\
&\quad \text{for} \quad V/V_y \le -2 + 2\sqrt{1/\alpha}
\end{aligned}$$

Secondly, the time-history response for CASE 2 is derived. Figures 6.6(b), (c) show the time histories of the deformation and the restoring force in CASE 2. The time-history response after the impulse acting point (point A in Figure 6.6) can be expressed as follows:

$$u(t) = \left\{\left(v_c + V\right)/\left(\sqrt{\alpha} V_y\right)\right\} d_y \sin\left(\sqrt{\alpha}\omega_1 t\right) - \left\{(1/\alpha) - 1\right\} d_y \tag{6.A9a}$$

$$\dot{u}(t) = (v_c + V)\cos\left(\sqrt{\alpha}\,\omega_1 t\right) \qquad (6.\text{A}9\text{b})$$

In Eqs. (6.A9a) and (6.A9b), $t = 0$ is set at point A. The parameters v_c, u_p can be obtained from Eqs. (6.13) and (6.15). The time interval between point A and B in Figure 6.6 is denoted by t_{AB} as shown in Figures 6.6(b), (c). t_{AB} can then be obtained as follows from $\dot{u}(t = t_{AB}) = 0$ and Eq. (6.A9b).

$$t_{AB}/T_1 = \{1/(2\pi)\}\{\pi/(2\sqrt{\alpha})\} = 1/(4\sqrt{\alpha}) \qquad (6.\text{A}10)$$

The time-history response after the unloading starting point (point B in Figure 6.6) can be expressed as follows:

$$u(t) = \{1 + 0.5\alpha(u_p/d_y)\}d_y\cos(\omega_1 t) + (1-\alpha)0.5u_p \qquad (6.\text{A}11\text{a})$$

$$\dot{u}(t) = -\{1 + 0.5\alpha(u_p/d_y)\}V_y\sin(\omega_1 t) \qquad (6.\text{A}11\text{b})$$

In Eqs. (6.A11a) and (6.A11b), $t = 0$ is set at point B. The time interval between point B and C in Figure 6.6 is denoted by t_{BC} as shown in Figures 6.6(b), (c). The time t_{BC} can then be obtained as follows from $u(t = t_{BC}) = -d_y + 0.5u_p$ and Eq. (6.A11a).

$$t_{BC}/T_1 = (1/4) - \{1/(2\pi)\}\arcsin\left[\frac{0.5\alpha(u_p/d_y)-1}{0.5\alpha(u_p/d_y)+1}\right] \qquad (6.\text{A}12)$$

The time-history response after the yielding starting point (point C in Figure 6.6) can be expressed as follows:

$$u(t) = -\sqrt{\left(-\frac{1}{\alpha}+0.5\frac{u_p}{d_y}\right)^2 + \left(2\frac{u_p}{d_y}\right)}\,d_y$$
$$\times\sin\left\{\sqrt{\alpha}\,\omega_1 t - \arctan\frac{0.5(u_p/d_y)-(1/\alpha)}{\sqrt{2u_p/d_y}}\right\} + \left(\frac{1}{\alpha}-1\right)d_y \qquad (6.\text{A}13\text{a})$$

$$\dot{u}(t) = -\sqrt{\alpha\left\{\left(-\frac{1}{\alpha}+0.5\frac{u_p}{d_y}\right)^2 + \left(2\frac{u_p}{d_y}\right)\right\}}$$
$$\times V_y\cos\left\{\sqrt{\alpha}\,\omega_1 t - \arctan\frac{0.5(u_p/d_y)-(1/\alpha)}{\sqrt{2u_p/d_y}}\right\} \qquad (6.\text{A}13\text{b})$$

In Eqs. (6.A13a) and (6.A13b), $t = 0$ is set at point C. The time interval between point C and D in Figure 6.6 is denoted by t_{CD} as shown in

Figures 6.6(b), (c). The time t_{CD} can then be obtained as follows from $u(t = t_{CD}) = \{(1/\alpha) - 1\}d_y$ and Eq. (6.A13a).

$$t_{CD}/T_1 = \frac{1}{\sqrt{\alpha}} \frac{1}{2\pi} \arctan\left\{\frac{0.5(u_p/d_y)-(1/\alpha)}{\sqrt{2(u_p/d_y)}}\right\} \qquad (6.A14)$$

From Eqs. (6.A10), (6.A12), and (6.A14) and Figures 6.6(b), (c), the time interval t_0^c between two consecutive impulses acting at the zero-restoring-force points (points A and D) in CASE 2 can then be obtained as follows:

$$\begin{aligned}
t_0^c/T_1 &= (t_{AB}/T_1) + (t_{BC}/T_1) + (t_{CD}/T_1) \\
&= \frac{1}{4}\left(1 + \frac{1}{\sqrt{\alpha}}\right) + \frac{1}{2\pi}\left[-\arcsin\left\{\frac{0.5\alpha(u_p/d_y)-1}{0.5\alpha(u_p/d_y)+1}\right\}\right. \\
&\quad + \frac{1}{\sqrt{\alpha}}\arctan\left\{\frac{0.5(u_p/d_y)-(1/\alpha)}{\sqrt{2(u_p/d_y)}}\right\}\right] \qquad (6.A15) \\
&\text{for } V/V_y > -2 + 2\sqrt{1/\alpha}
\end{aligned}$$

APPENDIX 2: ADJUSTMENT OF INPUT LEVEL OF MULTI-IMPULSE AND CORRESPONDING SINUSOIDAL WAVE

The adjustment method of input level of the multi-impulse and the corresponding sinusoidal wave is explained here based on the equivalence of the maximum Fourier amplitude. It is noted that although two different multi-impulses with modification in the first impulse are treated in Chapter 4 and in this chapter, the resulting relation of the multi-impulse and the multi-cycle sinusoidal waves is the same.

Consider the multi-impulse as a representative of a long-duration ground acceleration as shown in Figure 6.2(a), expressed by

$$\begin{aligned}
\ddot{u}_g(t) &= V\delta(t) - V\delta(t-t_0) + V\delta(t-2t_0) - V\delta(t-3t_0) + \cdots \\
&\quad + (-1)^{N-1}V\delta\{t-(N-1)t_0\} \qquad (6.A16)
\end{aligned}$$

where N is the number of impulses. The corresponding multi-cycle sinusoidal wave $\ddot{u}_g^{SW}(t)$ is expressed as follows:

$$\ddot{u}_g^{SW}(t) = A_l \sin(\omega_l t) \quad (0 \leq t \leq 0.5NT_l = Nt_0) \qquad (6.A17)$$

where A_l is the acceleration amplitude, $T_l = 2t_0$ is the excitation period, $\omega_l = 2\pi/T_l$ is the excitation circular frequency, and $V_l = A_l/\omega_l$ is the velocity amplitude. The number of cycles is half the number of impulses.

The maximum Fourier amplitude of the multi-impulse $\ddot{u}_g(t)$ and that of the corresponding multi-cycle sinusoidal wave $\ddot{u}_g^{SW}(t)$ can be derived as follows:

$$\max\left|\ddot{U}_g(\omega)\right| = V\left\{\max\left|\sum_{n=0}^{N-1}(-1)^n e^{-i\omega n t_0}\right|\right\} = NV \tag{6.A18}$$

$$\max\left|\ddot{U}_g^{SW}(\omega)\right| = A_l\left\{\max\left|\frac{2\pi t_0}{\pi^2 - (\omega t_0)^2}\sin(0.5N\omega t_0)\right|\right\} \tag{6.A19}$$

The function $f(\omega t_0) = 2\pi t_0 \,|\sin(0.5N\omega t_0)/\{\pi^2 - (\omega t_0)^2\}|$ can be defined from Eq. (6.A19). If N is sufficiently large (e.g., over 20 impulses), the function $f(x = \omega t_0)$ is maximized at $\omega t_0 = \pi$ and the maximum value of $f(x = \omega t_0)$ can be obtained as follows by using l'Hospital's theorem.

$$\lim_{\omega t_0 \to \pi}\left|\frac{\sin(0.5N\omega t_0)}{\pi^2 - (\omega t_0)^2}\right| = \lim_{\omega t_0 \to \pi}\left|\frac{0.5N\cos(0.5N\omega t_0)}{-2(\omega t_0)}\right| = \frac{N}{4\pi} \tag{6.A20}$$

N is assumed here to be an even number. From Eqs. (6.A18), (6.A19), and (6.A20), the following relation can be obtained by using the equivalence $\max\left|\ddot{U}_g(\omega)\right| = \max\left|\ddot{U}_g^{SW}(\omega)\right|$ in the maximum Fourier amplitude.

$$V_l = A_l/\omega_l = (2/\pi)V \tag{6.A21}$$

Chapter 7

Critical earthquake response of an elastic–perfectly plastic SDOF model on compliant ground under double impulse

7.1 INTRODUCTION

As in Chapters 2 and 5, the double impulse is introduced as a substitute of fling-step near-fault ground motion. It is well known that the ground beneath a structure influences the response of the structure during earthquake ground motions. It is therefore important and useful to include the effect of the ground on the responses of structures in the seismic-resistant design.

A closed-form expression for the elastic-plastic response of a structure on the compliant (flexible) ground by the "critical double impulse" is derived based on the expression for the corresponding structure with a fixed base. As in the case of the fixed-base model, only free-vibration appears under such double impulse, and the energy balance approach plays an important role in the derivation of the closed-form expression for a complicated elastic-plastic response on the compliant ground. It is remarkable that no iteration is needed in the derivation of the critical elastic-plastic response. It is shown via the closed-form expression that, in the case of a smaller input level of the double impulse to the structural strength, as the ground stiffness becomes larger, the maximum plastic deformation becomes larger. On the other hand, in the case of a larger input level of the double impulse to the structural strength, as the ground stiffness becomes smaller, the maximum plastic deformation becomes larger. The criticality and validity of the proposed theory are investigated through the comparison with the nonlinear time-history response analysis to the corresponding one-cycle sinusoidal input as a representative of the fling-step near-fault ground motion. The applicability of the proposed theory to actual recorded pulse-type ground motions is also discussed.

7.2 DOUBLE IMPULSE INPUT

7.2.1 Double impulse input

As explained in Kojima et al. (2015), Kojima and Takewaki (2015a, b), and Chapters 2 and 5, the fling-step input (fault-parallel) of the near-fault ground motion can be represented by a one-cycle sinusoidal wave, and the forward-directivity input (fault-normal) of the near-fault ground motion can be expressed by a series of three sinusoidal wavelets. The fling step is caused by the permanent displacement of the ground induced by the fault dislocation, and the forward directivity effect is concerned with the relation of the movement of the rupture front with the site. The discussion in this chapter is intended to simplify typical near-fault ground motions by a double impulse (Kojima et al. 2015, Kojima and Takewaki 2015a). This is because the double impulse has a simple characteristic, and a straightforward expression of the response can be expected even for elastic-plastic responses based on an energy approach to free vibrations. Furthermore, the double impulse enables us to describe directly the critical timing of impulses (resonant frequency), which is not easy for the sinusoidal and other inputs without application of a repetitive procedure. It is remarkable to note that while most of the previous methods (Caughey 1960a, b, Roberts and Spanos 1990, Luco 2014) employ the equivalent linearization of the structural model with the input unchanged (see Figure 7.1(a) including an equivalent linear stiffness), the method proposed in Kojima and Takewaki (2015a, b) and in this chapter transforms the input into the double impulse with the structural model unchanged (see Figure 7.1(b)).

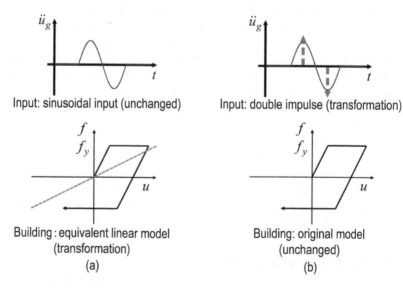

Figure 7.1 Comparison of present method with previous method: (a) previous method (equivalent linearization of structural model for unchanged input); (b) present method (transformation of input into double impulse for unchanged structural model) (Kojima and Takewaki 2016).

Consider again a ground acceleration $\ddot{u}_g(t)$ as double impulse expressed by

$$\ddot{u}_g(t) = V\delta(t) - V\delta(t - t_0) \tag{7.1}$$

where V is the given initial velocity, $\delta(t)$ is the Dirac delta function, and t_0 is the time interval between two impulses. It is well understood that the double impulse is a good approximation of the corresponding sinusoidal wave even in the form of velocity and displacement. However, the correspondence in the response should be discussed carefully. This will be conducted in Section 7.5.

As shown in Chapters 2 and 5, the Fourier transform of $\ddot{u}_g(t)$ of the double impulse can be derived as

$$\ddot{U}_g(\omega) = \int_{-\infty}^{\infty} \{V\delta(t) - V\delta(t - t_0)\} e^{-i\omega t} dt \tag{7.2}$$
$$= V\left(1 - e^{-i\omega t_0}\right)$$

7.2.2 Closed-form critical elastic-plastic response of SDOF system subjected to double impulse (summary of results in Chapter 2)

In Kojima and Takewaki (2015a) and Chapter 2, a closed-form expression for the critical response of an undamped elastic–perfectly plastic (EPP) SDOF system was derived for the double impulse (see Figure 7.2). The critical response plays a key role in the worst-case analysis (Drenick 1970, Takewaki 2002, 2007, Moustafa et al. 2010, Takewaki et al. 2012). Since this expression is used effectively in this chapter, the essence will be shown in this section.

Consider an undamped EPP SDOF system of mass m and stiffness k on the rigid ground. The yield deformation and force are denoted by d_y and f_y (see Figure 7.3). Let $\omega_1 = \sqrt{k/m}$, u, and f denote the undamped natural circular frequency, the displacement of the mass relative to the ground, and the restoring force of the model, respectively. The plastic deformation after the first impulse is expressed by u_{p1} and that after the second impulse is denoted by u_{p2}.

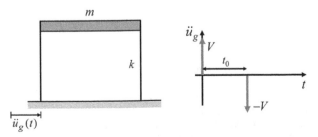

Figure 7.2 SDOF model subjected to critical (resonant) double impulse.

As explained before, the impulse input changes the mass velocity by V instantaneously and the elastic-plastic response of the EPP SDOF system under the double impulse can be expressed by the continuation of free vibrations. Let u_{max1} and u_{max2} denote the maximum deformations after the first and second impulses, respectively, as shown in Figure 7.3. Those responses can be derived by an energy balance approach without solving the equation of motion directly. The kinetic energies given at the initial stage by the first impulse and at the subsequent time by the second impulse are transformed into the sum of the hysteretic energy and the strain energy corresponding to the yield deformation (see Figures 1.5, 1.7). The critical timing of the second impulse corresponds to the state of the system with a zero restoring force, and only the kinetic energy exists in this state as mechanical energies. By using this rule, the maximum deformation can be obtained in a simple manner (see Figure 1.7).

The maximum elastic-plastic response of the undamped EPP SDOF system under the critical double impulse can be classified into the three cases depending on the yielding stage. The parameter $V_y(=\omega_1 d_y)$ denotes the input level of the velocity of the double impulse at which the undamped EPP SDOF system just attains the yield deformation after the first impulse. This parameter also indicates a strength parameter of the undamped EPP SDOF system. CASE 1 corresponds to the case of elastic response even after the second impulse and CASE 2 implies the case of plastic deformation only after the second impulse. Finally, CASE 3 presents the case of plastic deformation after the first impulse. Figure 7.3 shows the schematic diagram for these three cases.

Figure 7.3(a) shows the maximum deformation after the first impulse and that after the second impulse, respectively, for the elastic case (CASE 1) during the whole stage. The maximum deformations u_{max1} and u_{max2} can be obtained as follows from the energy balance.

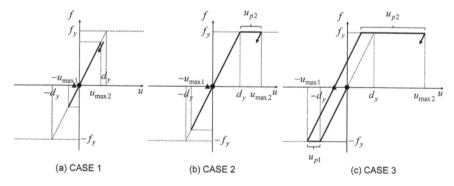

Figure 7.3 Prediction of maximum elastic-plastic deformation to double impulse based on energy approach: (a) CASE 1: Elastic response; (b) CASE 2: Plastic response after second impulse; (c) CASE 3: Plastic response after first impulse (● : first impulse, ▲ : second impulse) (Kojima and Takewaki 2016).

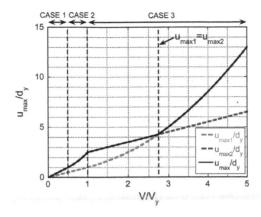

Figure 7.4 Maximum normalized elastic-plastic deformation under double impulse with respect to input level (Kojima and Takewaki 2015a).

$$u_{max1}/d_y = V/V_y \tag{7.3}$$

$$u_{max2}/d_y = 2(V/V_y) \tag{7.4}$$

A similar energy balance law enables the derivation of u_{max1} and u_{max2} for CASEs 2 and 3 (Figures 7.3(b), (c)).

$$u_{max1}/d_y = V/V_y \quad (\text{CASE 2}) \tag{7.5}$$

$$u_{max2}/d_y = 0.5\{1+(2V/V_y)^2\} \quad (\text{CASE 2}) \tag{7.6}$$

$$u_{max1}/d_y = 0.5\{1+(V/V_y)^2\} \quad (\text{CASE 3}) \tag{7.7}$$

$$u_{max2}/d_y = 0.5(3+2V/V_y) \quad (\text{CASE 3}) \tag{7.8}$$

Figure 7.4 shows the maximum value of normalized elastic-plastic deformation under the double impulse with respect to input level.

7.3 MAXIMUM ELASTIC-PLASTIC DEFORMATION OF SIMPLIFIED SWAYING-ROCKING MODEL TO CRITICAL DOUBLE IMPULSE

7.3.1 Simplified swaying-rocking model

In this chapter, a closed-form expression (Kojima and Takewaki 2015a) for the maximum elastic-plastic response of an undamped EPP SDOF system on

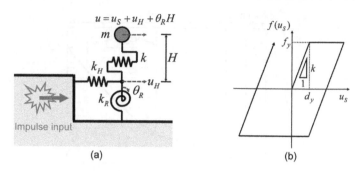

$$u = u_S + u_H + \theta_R H$$

Figure 7.5 Simplified swaying-rocking model: (a) model parameters, (b) restoring-force characteristic (Kojima and Takewaki 2016).

the rigid ground under the critical double impulse is used to derive the closed-form expression for the maximum elastic-plastic response of a simplified SR (Swaying-Rocking) model, as shown in Figure 7.5(a), under the critical double impulse. In the simplified SR model, the foundation mass and floor mass moments of inertia are neglected in the original SR model. Note that the critical impulse intervals of the undamped EPP SDOF system on the rigid ground and that on the flexible ground are different. More specifically, the latter is longer than the former one. This characteristic will be clarified in this chapter.

The superstructure is the same as explained in Section 7.2.2. Let k_H and k_R denote the swaying and rocking spring stiffnesses, respectively. The restoring-force characteristic of the superstructure is shown in Figure 7.5(b). The damping of the superstructure and the ground is neglected here for a simple explanation of the effect of the ground stiffness on the maximum elastic-plastic response of the superstructure. Actually, it is well known that the viscous damping is usually not effective for the impulsive loading like the near-fault ground motions. Let u_S, u_H, and θ_R denote the actual deformation of the super-structure, the swaying spring deformation and the angle of rotation of the rocking spring. H denotes the equivalent height of the super-structure mass.

In this chapter, the swaying and rocking spring stiffnesses are assumed to be expressed by

$$k_H = \{6.77/(1.97 - v)\} Gr \tag{7.9}$$

$$k_R = \{2.52/(1.00 - v)\} Gr^3 \tag{7.10}$$

where v, r, G, ρ, V_S denote the Poisson's ratio of the ground, the equivalent radius of the foundation (in the case of transformation of the footing to an equivalent circular disc), the shear modulus of the ground, the mass density of the ground, and the shear wave velocity of the ground (Parmelee 1970).

The equations of motion for the simplified SR model in the elastic range can be described by

$$m\left(\ddot{u}_S + \ddot{u}_H + \ddot{\theta}_R H\right) + ku_S = -m\ddot{u}_g \qquad (7.11a)$$

$$ku_S - k_H u_H = 0 \qquad (7.11b)$$

$$ku_S H - k_R \theta_R = 0 \qquad (7.11c)$$

On the other hand, the equations of motion for the simplified SR model in the elastic-plastic range can be expressed by

$$m\left(\ddot{u}_S + \ddot{u}_H + \ddot{\theta}_R H\right) + f\left(u_S\right) = -m\ddot{u}_g \qquad (7.12a)$$

$$f\left(u_S\right) - k_H u_H = 0 \qquad (7.12b)$$

$$\left\{f\left(u_S\right)\right\} H - k_R \theta_R = 0 \qquad (7.12c)$$

where $f(u_S)$ is the restoring force in the super-structure.

7.3.2 Equivalent SDOF model of simplified swaying-rocking model

In order to use the expression in Section 7.2.2, consider an equivalent SDOF model of the simplified swaying-rocking model as shown in Figure 7.6(a). The equation of motion of the equivalent SDOF model in the elastic range can be described by

$$m\ddot{u}^e + k^e u^e = -m\ddot{u}_g \qquad (7.13)$$

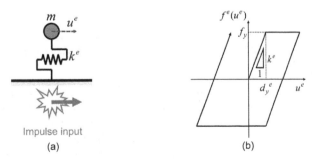

Figure 7.6 Equivalent SDOF model of simplified swaying-rocking model: (a) model parameters, (b) restoring-force characteristic (Kojima and Takewaki 2016).

where u^e and k^e are the displacement of mass and the elastic stiffness in the equivalent SDOF model. Since the three springs are connected in series, k^e can be expressed as follows in terms of the stiffnesses of the three springs (Veletsos and Meek 1974, Veletsos 1977, Jarernprasert et al. 2013).

$$k^e = k/\{1+(k/k_H)+(k/k_R)H^2\} \tag{7.14}$$

Because the yield force in the equivalent SDOF model is equal to the yield force of the superstructure, the yield displacement of the equivalent SDOF model can be expressed by

$$d_y^e = f_y/k^e = \{1+(k/k_H)+(k/k_R)H^2\}d_y \tag{7.15}$$

In addition, the natural frequency of the equivalent SDOF model is computed as

$$\omega_1^e = \sqrt{k^e/m} = \sqrt{1/\{1+(k/k_H)+(k/k_R)H^2\}}\omega_1 \tag{7.16}$$

Using Eqs. (7.15) and (7.16), the reference input level corresponding to $V_y(=\omega_1 d_y)$ for the superstructure can be defined by

$$\begin{aligned} V_y^e = \omega_1^e d_y^e &= \sqrt{\{1+(k/k_H)+(k/k_R)H^2\}}\omega_1 d_y \\ &= \sqrt{\{1+(k/k_H)+(k/k_R)H^2\}}V_y \end{aligned} \tag{7.17}$$

After the simplified SR model is transformed into the equivalent SDOF model defined in Figure 7.6, the maximum elastic-plastic response of the equivalent SDOF model under the critical double impulse can be obtained in the same manner as shown in Figure 7.7 by replacing the parameters with the equivalent parameters.

7.3.3 Critical elastic-plastic response of simplified swaying-rocking model subjected to double impulse

In this section, the maximum elastic-plastic response is derived for the superstructure in the simplified SR model under the critical double impulse. As mentioned above, the critical impulse intervals of the undamped EPP SDOF system on the rigid ground and that on the flexible ground are different. The latter is longer than the former. The ductility factor of the superstructure can be expressed by $(d_y + u_p)/d_y$. Although the ductility factor of the superstructure can be derived from the ductility factor of the equivalent SDOF model introduced in Section 7.3.2 by modifying the coefficient, the

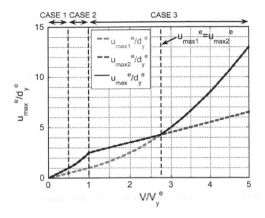

Figure 7.7 Maximum deformation of equivalent SDOF model with respect to input level normalized for equivalent SDOF model (Kojima and Takewaki 2016).

direct expression of the ductility factor of the superstructure is obtained here.

As in Section 7.2.2, three cases, CASE 1, CASE 2, and CASE 3, are treated depending on the input level of the double impulse.

First of all, consider CASE 1 (see Figure 7.8(a)). The energy balances after the first impulse and the second impulse are described as follows.

$$mV^2/2 = ku_{S\max1}^2/2 + k_H u_{H1}^2/2 + k_R \theta_{R1}^2/2 \qquad (7.18)$$

$$m(2V)^2/2 = ku_{S\max2}^2/2 + k_H u_{H2}^2/2 + k_R \theta_{R2}^2/2 \qquad (7.19)$$

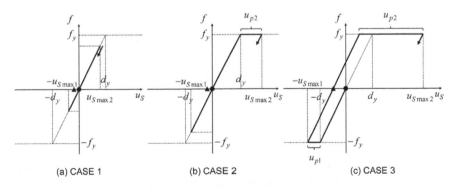

(a) CASE 1   (b) CASE 2   (c) CASE 3

Figure 7.8 Prediction of maximum elastic-plastic deformation of superstructure on compliant ground to double impulse based on energy approach: (a) Case 1: Elastic response; (b) Case 2: Plastic response after second impulse; (c) Case 3: Plastic response after first impulse (●: first impulse, ▲: second impulse) (Kojima and Takewaki 2016).

where $u_{S\,max\,1}$, u_{H1}, and θ_{R1} denote the maximum deformations of the super-structure, the swaying spring, and the rocking spring, respectively, after the first impulse. In addition, $u_{S\,max\,2}$, u_{H2}, and θ_{R2} denote those after the second impulse. The displacements u_{H1}, θ_{R1}, u_{H2}, and θ_{R2} in the elastic range can be expressed as follows in terms of $u_{S\,max\,1}$ and $u_{S\,max\,2}$.

$$u_{H1} = (k/k_H)u_{S\,max1}, \quad \theta_{R1} = (kH/k_R)u_{S\,max1} \qquad (7.20\text{a, b})$$

$$u_{H2} = (k/k_H)u_{S\,max2}, \quad \theta_{R2} = (kH/k_R)u_{S\,max2} \qquad (7.21\text{a, b})$$

Substitution of Eqs. (7.20a, b) and (7.21a, b) into Eqs. (7.18) and (7.19) and rearrangement of the resulting equations provide

$$u_{S\,max1} = \sqrt{1/\left\{1+(k/k_H)+\left(kH^2/k_R\right)\right\}}\,(V/\omega_1) \qquad (7.22)$$

$$u_{S\,max2} = \sqrt{1/\left\{1+(k/k_H)+\left(kH^2/k_R\right)\right\}}\,(2V/\omega_1) \qquad (7.23)$$

Dividing both sides of Eqs. (7.22) and (7.23) by d_y and taking into account the definition $V_y = \omega_1 d_y$, the following expressions can be derived.

$$u_{S\,max1}/d_y = \sqrt{1/\left\{1+(k/k_H)+\left(kH^2/k_R\right)\right\}}\,(V/V_y) \qquad (7.24)$$

$$u_{S\,max2}/d_y = \sqrt{1/\left\{1+(k/k_H)+\left(kH^2/k_R\right)\right\}}\,(2V/V_y) \qquad (7.25)$$

The parameter V_y is the reference input level for the fixed-base superstructure introduced in Section 7.2.2 and indicates the strength of the superstructure in terms of input velocity.

Figure 7.8(b) presents the elastic-plastic deformation of the superstructure after the second impulse in CASE 2 (the superstructure goes into the plastic range after the second impulse). From Eq. (7.25), the condition that the superstructure goes into the plastic range after the second impulse can be described by

$$V/V_y > 0.5\sqrt{1+(k/k_H)+\left(kH^2/k_R\right)} \qquad (7.26)$$

In CASE 2, the superstructure is in the elastic range in the process between the first impulse and the second impulse. The energy balance after the second impulse can be described as follows by using the energy balance law and Figure 7.8(b).

$$m(2V)^2/2 = f_y d_y/2 + f_y u_{p2} + k_H u_{H2}^2/2 + k_R \theta_{R2}^2/2 \qquad (7.27)$$

where u_{p2}, u_{H2}, θ_{R2} are the maximum plastic deformation of the superstructure and the maximum deformations of the swaying and rocking springs. The restoring force of the superstructure in the plastic range after the second impulse is described by

$$f(u_S) = f_y \tag{7.28}$$

The displacements u_{H2}, θ_{R2} can be derived from Eqs. (7.12b, c) and (7.28).

$$u_{H2} = f_y / k_H = (k/k_H) d_y, \quad \theta_{R2} = (f_y H)/k_R = (kH/k_R) d_y \tag{7.29a, b}$$

Then, the plastic deformation u_{p2} after the second impulse can be obtained by substituting Eqs. (7.29a, b) into Eq. (7.27) and rearranging the resulting equation.

$$u_{p2} = 0.5 \left[\left\{ (2V)/(\omega_1 d_y) \right\}^2 - \left\{ 1 + (k/k_H) + (kH^2/k_R) \right\} \right] d_y \tag{7.30}$$

Dividing both sides of Eq. (7.30) by d_y and recalling the definition $\omega_1 d_y = V_y$, the following relation can be derived.

$$u_{p2}/d_y = 0.5 \left[\left\{ (2V)/V_y \right\}^2 - \left\{ 1 + (k/k_H) + (kH^2/k_R) \right\} \right] \tag{7.31}$$

Figure 7.8(c) presents the plastic deformations after the first and second impulses in CASE 3 (the superstructure goes into the plastic range after the first impulse). The condition that the superstructure goes into the plastic range after the first impulse can be derived as follows from Eq. (7.24).

$$V/V_y > \sqrt{1 + (k/k_H) + (kH^2/k_R)} \tag{7.32}$$

Using the energy balance law and Figure 7.8(c), the energy balances after the first and second impulses can be described by

$$mV^2/2 = f_y d_y/2 + f_y u_{p1} + k_H u_{H1}^2/2 + k_R \theta_{R1}^2/2 \tag{7.33}$$

$$m(v_c + V)^2/2 = f_y d_y/2 + f_y u_{p2} + k_H u_{H2}^2/2 + k_R \theta_{R2}^2/2 \tag{7.34}$$

where u_{p1}, u_{H1}, θ_{R1} denote the maximum plastic deformation of the superstructure and the maximum deformations of the swaying and rocking springs after the first impulse. In addition, u_{p2}, u_{H2}, θ_{R2} are those after the second impulse. The restoring-force in the superstructure can be expressed as

$$f(u_S) = -f_y \tag{7.35}$$

The displacements u_{H1}, θ_{R1} can be obtained from Eqs. (7.12b, c) and (7.35).

$$u_{H1} = -f_y/k_H = -(k/k_H)d_y,$$
$$\theta_{R1} = -(f_y H)/k_R = -(kH/k_R)d_y \qquad \text{(7.36a, b)}$$

The displacements u_{H2}, θ_{R2} in CASE 3 are the same as Eqs. (7.29a, b). The parameter v_c in Eq. (7.34) is the superstructure mass velocity when the restoring-force of the superstructure becomes zero after the first impulse. The energy balance after the initiation of unloading before the second impulse provides

$$f_y d_y/2 + k_H u_{H1}^2/2 + k_R \theta_{R1}^2/2 = m v_c^2/2 \qquad \text{(7.37)}$$

Substitution of Eqs. (7.36a, b) into Eq. (7.37) and rearrangement of the resulting equation lead to the following expression of v_c.

$$v_c = \sqrt{1 + (k/k_H) + (kH^2/k_R)}\,(\omega_1 d_y)$$
$$= \sqrt{1 + (k/k_H) + (kH^2/k_R)}\,V_y \qquad \text{(7.38)}$$

Substituting Eqs. (7.29a, b) and (7.36a, b) into Eqs. (7.33) and (7.34) and rearranging the resulting equations, the following expressions for the plastic deformations can be drawn.

$$u_{p1} = 0.5\left[\{V/(\omega_1 d_y)\}^2 - \{1 + (kH^2/k_R) + (k/k_H)\}\right]d_y \qquad \text{(7.39)}$$

$$u_{p2} = 0.5\left[\{(v_c + V)/(\omega_1 d_y)\}^2 - \{1 + (kH^2/k_R) + (k/k_H)\}\right]d_y \qquad \text{(7.40)}$$

Dividing both sides of Eqs. (7.39) and (7.40) by d_y and using Eq. (7.38) and the definition $\omega_1 d_y = V_y$, the following relations can be derived.

$$u_{p1}/d_y = 0.5\left[(V/V_y)^2 - \{1 + (kH^2/k_R) + (k/k_H)\}\right] \qquad \text{(7.41)}$$

$$u_{p2}/d_y = 0.5(V/V_y)^2 + \sqrt{1 + (kH^2/k_R) + (k/k_H)}\,(V/V_y) \qquad \text{(7.42)}$$

After the comparison of Eqs. (7.41) and (7.42), $u_{p2}/d_y > u_{p1}/d_y$ can be confirmed. Therefore, we use $u_p/d_y = u_{p2}/d_y$. Finally, u_p/d_y can be derived as follows.

$$u_p/d_y = \begin{cases} 0.5\left\{\left(\dfrac{2V}{V_y}\right)^2 - \left(1 + \dfrac{k}{k_H} + \dfrac{kH^2}{k_R}\right)\right\} & \text{for } 0.5\sqrt{1 + \dfrac{k}{k_H} + \dfrac{kH^2}{k_R}} \\ & \leq V/V_y < \sqrt{1 + \dfrac{k}{k_H} + \dfrac{kH^2}{k_R}} \\ 0.5\left(\dfrac{V}{V_y}\right)^2 + \sqrt{1 + \dfrac{k}{k_H} + \dfrac{kH^2}{k_R}}\left(\dfrac{V}{V_y}\right) & \text{for } \sqrt{1 + \dfrac{k}{k_H} + \dfrac{kH^2}{k_R}} \leq V/V_y \end{cases}$$

$$(7.43)$$

7.3.4 Numerical example

In this section, the effect of soil types on the response of superstructures is investigated using the closed-form expressions derived in the previous section.

Consider three soil conditions, i.e., soil type 1, 2, 3. The shear wave velocities of soil type 1, 2, 3 are $V_S = 200$[m/s] for soil type 1 (stiff), $V_S = 133$[m/s] for soil type 2 (medium), and $V_S = 100$[m/s] for soil type 3 (soft). The mass of the SDOF superstructure is $m = 800 \times 10^3$[kg] and the natural period of the superstructure with fixed-base is 1.0[s]. The superstructure is to be modeled from a 10-story building. The yield deformation is $d_y = 0.16$[m], the equivalent height is $H = 40 \times 0.7 = 28$[m], and the equivalent radius of the foundation is $r = 8$[m]. The mass density of ground is $\rho = 1.8 \times 10^3$[kg/m³] and the Poisson's ratio is $\nu = 0.35$.

Figure 7.9 shows the relation of ductility factor $(d_y + u_p)/d_y$ with V/V_y for three soil conditions and fixed-base case. V is the initial velocity of the

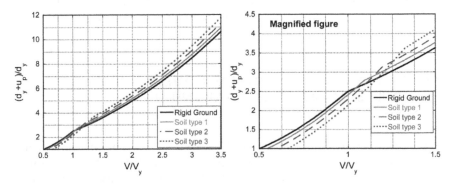

Figure 7.9 Relation of maximum plastic deformation $(d_y + u_p)/d_y$ with V/V_y for three soil conditions and fixed-base case (Kojima and Takewaki 2016).

double impulse. In the present numerical example, $(d_y + u_p)/d_y$ derived from Eq. (7.43) is treated and the input level is normalized for $V_y = \omega_1 d_y$ of the superstructure with a fixed base.

Some observations can be drawn from Figure 7.9. In CASE 2 (low input level), as the ground becomes stiffer, the plastic deformation of the super-structure becomes larger. On the other hand, in CASE 3 (large input level), as the ground becomes softer, the plastic deformation of the superstructure becomes larger. These properties may come from the fact that, as the ground becomes softer, the strain energy stored in the ground becomes larger in the case where the super-structure is in the plastic range.

In more detail, in CASE 2, the input energy (the left-hand side of Eq. (7.27)) at the second impulse is constant. As the ground becomes stiffer, the strain energy (the third and fourth terms of the right-hand side of Eq. (7.27)) stored in the ground becomes smaller and the plastic deformation (the second term of the right-hand side of Eq. (7.27)) becomes larger (see Figure 7.10(a)).

On the other hand, in CASE 3, the mass velocity (the right-hand side of Eq. (7.37) and Eq. (7.38)) just before the second impulse becomes larger from Eq. (7.37) as the ground becomes softer (as the strain energy (the second and third terms of the left-hand side of Eq. (7.37)) stored in the ground becomes larger). Then the input energy (the left-hand side of Eq. (7.34)) at the second impulse becomes larger and the plastic deformation after the second impulse becomes larger (see Figure 7.10(b)). In discussing this phenomenon, note that the velocity is included in the form of second order in Eq. (7.34). This result can also be understood from the difference in the second terms in Eq. (7.43). The closed-form expression in Eq. (7.43) is very useful in the clarification of the effect of soil types on the superstructure response. In the present formulation, the damping of the ground was disregarded. Since the damping is larger in the soft ground, the values in Figure 7.9 become smaller for softer ground in general. However, it is also true that the ground damping is less effective under impulsive inputs. These influences will be studied in the future.

7.4 APPLICABILITY OF CRITICAL DOUBLE IMPULSE TIMING TO CORRESPONDING SINUSOIDAL WAVE

In the reference (Kojima and Takewaki 2015a), it was demonstrated that if the maximum value of the Fourier amplitude is selected as the key parameter for an adjustment (Section 1.2 in Chapter 1), the responses to the double impulse and the corresponding sinusoidal input exhibit a fairly good correspondence. In this section, it is investigated whether the critical timing derived from the double impulse is also an approximate critical timing of the sinusoidal input. Although the SDOF model with a fixed base is treated

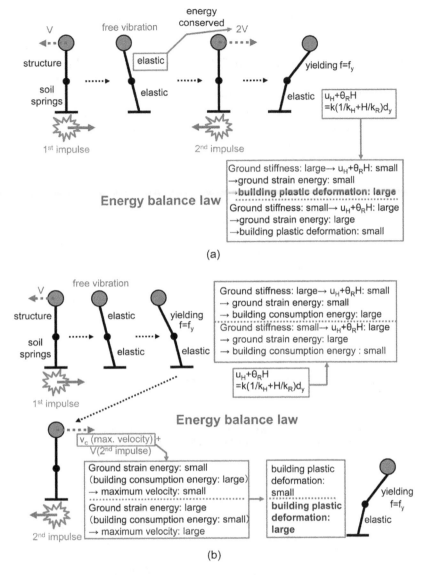

Figure 7.10 Schematic diagram for better understanding of effect of soil type on structure plastic deformation: (a) small input level (Case 2), (b) large input level (Case 3) (Kojima and Takewaki 2016).

here, it is applicable to the equivalent SDOF model by introducing the equivalent parameters, V_y^e, d_y^e, ω_1^e etc.

Let t_0^c denote the critical timing of the double impulse and t_0 denote the general timing. In Kojima and Takewaki (2015a) and Chapter 2, t_0^c has been derived as follows (CASE 3: large input level; plastic deformation after the first impulse).

$$t_0^c / T_1 = \left\{ \arcsin\left(V_y / V \right) \right\} / \left(2\pi \right) + \left\{ \sqrt{\left(V / V_y \right)^2 - 1} / \left(2\pi \right) \right\} + 1/4 \qquad (7.44)$$

This relation is plotted in Figure 7.11. It can be understood that the critical timing is delayed due to plastic deformation as the input level increases.

Figure 7.12(a) shows the maximum deformation with respect to t_0 / t_0^c. It can be observed that the critical time interval t_0^c derived from the double impulse is a good approximate of that for the sinusoidal input. Figure 7.12(b) is the corresponding plot for the double impulse (Kojima and Takewaki 2015a).

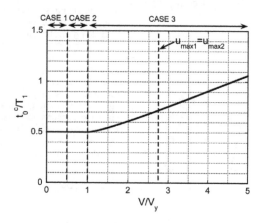

Figure 7.11 Interval time between first and second impulses with respect to input level (Kojima and Takewaki 2015a).

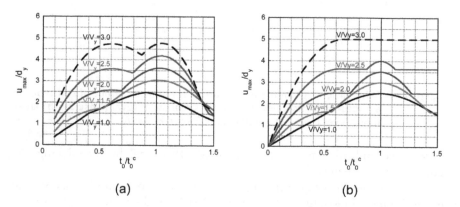

(a)   (b)

Figure 7.12 Maximum deformation with respect to t_0 / t_0^c: (a) sinusoidal input, (b) double impulse (Kojima and Takewaki 2016).

7.5 TOWARD BETTER CORRESPONDENCE BETWEEN DOUBLE IMPULSE AND SINUSOIDAL INPUT

In Kojima and Takewaki (2015a) and Chapter 2, it was clarified that if the magnitude of the double impulse is adjusted so that the maximum values of the Fourier amplitudes of the double impulse and the corresponding sinusoidal input are the same (Section 1.2 in Chapter 1), the maximum elastic-plastic responses correspond well in the range of input level $V/V_y < 3$. However, in the range of $V/V_y > 3$, the maximum response of the double impulse becomes larger than that of the sinusoidal input. In order to seek a better correspondence over a wider range of input level, the amplitude of the sinusoidal input is amplified. This is because the effect of the sinusoidal input is rather small in the large input level $V/V_y > 3$ compared to the large influence of the first impulse in the double impulse. As in Section 7.4, although the SDOF model with a fixed base is treated here, it is applicable to the equivalent SDOF model by introducing the equivalent parameters, V_y^e, d_y^e, ω_1^e etc.

Figure 7.13(a) presents the plot of the coefficient a with respect to the timing of the double impulse for adjusting the maximum Fourier amplitudes of the double impulse and the sinusoidal input where the sinusoidal acceleration input is expressed as $\ddot{u}_g(t) = A\sin(\pi t/t_0)$ and the coefficient a is defined by $a = A/V$. Figures 7.13(b)–(e) present the maximum normalized elastic-plastic deformations to the double impulse and the corresponding sinusoidal inputs amplified by 1.0, 1.1, 1.15, 1.2 from the original input with the same maximum Fourier amplitude as the double impulse. It can be found that the amplification 1.15 or 1.2 provide the best fitting for the input range of $V/V_y > 3$.

7.6 APPLICABILITY TO RECORDED GROUND MOTIONS

Because the double impulse and the corresponding one-cycle sinusoidal wave have somewhat different natures from actual recorded ground motions, it seems important to investigate the applicability of the present theory to actual recorded pulse-type ground motions. As explained in the previous sections, the maximum deformation of the simplified SR model can be obtained from the equivalent SDOF model. Figure 7.7 enables the estimation of the maximum deformation of the equivalent SDOF model. Therefore, it is sufficient to investigate the response of the SDOF model. Moreover, since the consideration and reflection of ground conditions in actual recorded ground motions is complicated, an SDOF model with a fixed base is considered in this section.

Consider two pulse-type recorded ground motions, the Rinaldi station fault-normal (FN) component during the Northridge earthquake in 1994

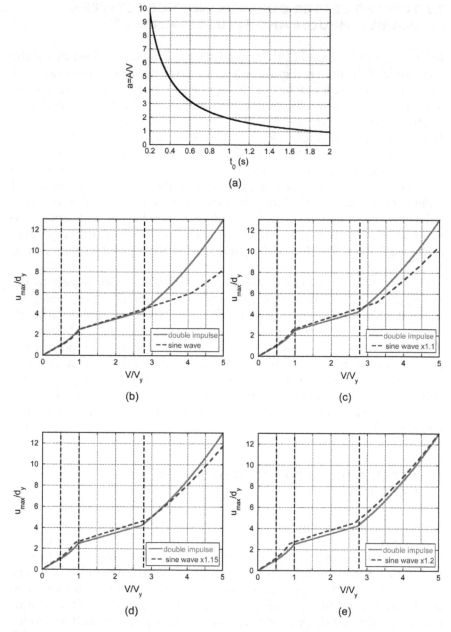

Figure 7.13 Better correspondence of maximum responses to double impulse and amplified sinusoidal inputs: (a) plot of coefficient a with respect to timing of double impulse; (b) amplification factor = 1.0; (c) amplification factor = 1.1; (d) amplification factor = 1.15; (e) amplification factor = 1.2 (Kojima and Takewaki 2016).

and the Kobe University NS component (almost fault-normal) during the Hyogoken-Nanbu (Kobe) earthquake in 1995. The input level of the equivalent double impulse for the Rinaldi station FN component is $V = 1.64$ [m/sec] and that for the Kobe University NS component is $V = 0.677$ [m/sec]. As in the previous chapter (Chapter 5), since the ground motions are fixed, the structural models are varied for the realization of the resonance, i.e., ω_1 or d_y in $V_y = \omega_1 d_y$ is varied. Figure 7.14 illustrates the modeling of the part of the recorded ground motion acceleration into a one-cycle sinusoidal input. Figure 7.15 shows the maximum amplitude of deformation for the recorded ground motions and the corresponding proposed one. As stated before, since the initial velocity V is determined in Figure 7.14, V_y is changed here.

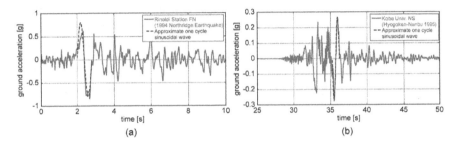

Figure 7.14 Modeling of part of pulse-type recorded ground motion into corresponding one-cycle sinusoidal input: (a) Rinaldi station fault-normal component during the Northridge earthquake in 1994; (b) Kobe University NS component (almost fault-normal) during the Hyogoken-Nanbu (Kobe) earthquake in 1995 (Kojima and Takewaki 2016).

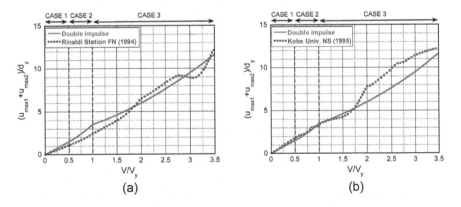

Figure 7.15 Maximum amplitude of deformation for recorded ground motions and proposed one: (a) Rinaldi station fault-normal component; (b) Kobe University NS component (Kojima and Takewaki 2016).

Because ω_1 is closely related to the resonance condition, d_y is changed principally. This procedure is similar to the well-known elastic-plastic response spectrum developed in 1960–1970. The solid line is obtained by changing V_y for the specified V using the method for the double impulse and the dotted line is drawn by conducting the elastic-plastic time-history response analysis on each model with varied V_y under the recorded ground motion. It can be observed that the result by the proposed method is a fairly good approximate of the recorded pulse-type ground motions.

7.7 SUMMARIES

The double impulse was introduced as a substitute for the fling-step near-fault ground motion. A closed-form expression for the elastic-plastic response of an elastic–perfectly plastic (EPP) single-degree-of-freedom (SDOF) model on the compliant (flexible) ground by the "critical double impulse" was derived based on the expression for the corresponding EPP SDOF model with a fixed base. The obtained results may be summarized as follows:

1. The closed-form expression for the maximum response of an EPP SDOF model with a fixed base under the critical double impulse was extended to a simplified swaying-rocking (SR) model. The simplified SR model can be derived by neglecting the foundation mass and floor mass moments of inertia in the original SR model.
2. The simplified SR model was transformed into an equivalent SDOF model with only one equivalent spring which was modeled from three springs (super-structure, swaying and rocking) via static condensation. By applying the work for an EPP SDOF model (Kojima and Takewaki 2015a, Chapter 2 in this book) to the equivalent SDOF model, the closed-form expression for the maximum elastic-plastic response of the equivalent SDOF model under the critical double impulse was derived and the corresponding elastic-plastic deformation of the superstructure was evaluated. No iteration is required in its derivation.
3. The closed-form expression for the critical elastic-plastic response of the superstructure enabled the clarification of the relation of the critical elastic-plastic response of the superstructure with the ground stiffness. In the case of a smaller input level of the double impulse to the structural strength, as the ground stiffness becomes larger, the maximum plastic deformation becomes larger. On the other hand, in the case of a larger input level of the double impulse to the structural strength, as the ground stiffness becomes smaller, the maximum plastic deformation becomes larger. This property can be explained from

the viewpoint of the elastic strain energy stored in the swaying and rocking springs. In the smaller input level (Case 2), as the ground stiffness becomes larger, the ground deformation after the second impulse becomes smaller and the elastic strain energy stored in the swaying and rocking springs becomes smaller. Then the plastic deformation of the superstructure becomes larger. On the other hand, in the larger input level (Case 3), as the ground stiffness becomes smaller, the elastic strain energy stored in the swaying and rocking springs during the plastic deformation in the superstructure after the first impulse becomes larger. This elastic strain energy stored in the swaying and rocking springs during the plastic deformation in the superstructure plays an important role in the magnitude of plastic deformation of the superstructure after the second impulse.

4. The critical timing derived from the double impulse is also an approximate critical timing of the sinusoidal input.
5. In the input range of $V/V_y > 3$, the maximum response of the double impulse becomes larger than that of the sinusoidal input. The better correspondence over a wider input range can be achieved by amplifying the amplitude of the sinusoidal input. The amplification factor 1.15 or 1.2 provides the best fitting for the input range of $V/V_y > 3$.
6. The method using the double impulse is applicable to actual recorded pulse-type ground motions within reasonable accuracy.

The property in item (3) indicates that when we design a building on a soft ground, we have to design stronger members in such a building. On the contrary, it is generally believed that the soil-structure interaction can reduce the building response under earthquake ground motion. The period of pulse-like waves is in the range of 0.5–3 sec. Since the critical input means the resonant case, the present theory dealing with the resonant response should be applied to buildings except very flexible ones of which the fundamental natural period is longer than 3 sec.

In this chapter, an explicit critical response of the EPP SDOF model on the compliant ground under the double impulse was derived. It is expected that the formulation is applied to the critical response under the triple impulse explained in Chapter 3 (Kojima and Takewaki 2015b).

REFERENCES

Caughey, T. K. (1960a). Sinusoidal excitation of a system with bilinear hysteresis, *J. Appl. Mech.*, 27(4), 640–643.
Caughey, T. K. (1960b). Random excitation of a system with bilinear hysteresis, *J. Appl. Mech.*, 27(4), 649–652.
Drenick, R. F. (1970). Model-free design of aseismic structures, *J. Eng. Mech. Div.*, ASCE, 96(EM4), 483–493.

Jarernprasert, S., Bazan, E., and Bielak, J. (2013). Seismic soil-structure interaction response of inelastic structures, *Soil Dyn. Earthquake Eng.*, 47, 132–143.

Kojima, K., Fujita, K., and Takewaki, I. (2015). Critical double impulse input and bound of earthquake input energy to building structure, *Frontiers in Built Environment*, 1: 5.

Kojima, K., and Takewaki, I. (2015a). Critical earthquake response of elastic-plastic structures under near-fault ground motions (Part 1: Fling-step input), *Frontiers in Built Environment*, 1: 12.

Kojima, K., and Takewaki, I. (2015b). Critical earthquake response of elastic-plastic structures under near-fault ground motions (Part 2: Forward-directivity input), *Frontiers in Built Environment*, 1: 13.

Kojima, K., and Takewaki, I. (2016). Closed-form critical earthquake response of elastic-plastic structures on compliant ground under near-fault ground motions, *Frontiers in Built Environment*, 2: 1.

Moustafa, A., Ueno, K., and Takewaki, I. (2010). Critical earthquake loads for SDOF inelastic structures considering evolution of seismic waves, *Earthquakes and Structures*; 1(2): 147–162.

Parmelee, R. A. (1970). The influence of foundation parameters on the seismic response of interaction systems, in *Proceedings of the 3rd Japan earthquake engineering symposium*, 3: 49–56.

Roberts, J. B., and Spanos, P. D. (1990). *Random vibration and statistical linearization*, Wiley, New York.

Takewaki, I. (2002). Robust building stiffness design for variable critical excitations, *J. Struct. Eng.*, ASCE, 128(12): 1565–1574.

Takewaki, I. (2007). *Critical excitation methods in earthquake engineering*, Elsevier, Amsterdam, Second edition in 2013, London.

Takewaki, I., Moustafa, A., and Fujita, K. (2012). *Improving the earthquake resilience of buildings: The worst case approach*, Springer, London.

Veletsos, A. S., and Meek, J.W. (1974). Dynamic behaviour of building-foundation systems, *Earthquake Eng. Struct. Dyn.*, 3: 121–138.

Veletsos, A. S. (1977). Dynamics of structure-foundation systems, in *Structural and geotechnical mechanics, a volume honoring N. M. Newmark*, edited by W. J. Hall, pp. 333–361, Englewood Cliffs, NJ: Prentice-Hall.

Chapter 8

Closed-form dynamic collapse criterion for a bilinear hysteretic SDOF model under near-fault ground motions

8.1 INTRODUCTION

Since the principal objective in the field of earthquake and structural engineering is how to design safe and reliable structures, the clarification of the phenomenon of dynamic instability or dynamic collapse is of principal interest (see Jennings and Husid 1968, Sun et al. 1973, Tanabashi et al. 1973, Bertero et al. 1978, Takizawa and Jennings 1980, Bernal 1987, 1998, Nakajima et al. 1990, Ger et al. 1993, Challa and Hall 1994, Uetani and Tagawa 1998, Hall 1998, Hjelmstad and Williamson 1998, Sasani and Bertero 2000, Araki and Hjelmstad 2000, Ibarra and Krawinkler 2005, Adam and Jager 2012). The theoretical investigations on the phenomenon of dynamic instability or dynamic collapse from the viewpoint of applied mechanics have also been conducted extensively (Herrmann 1965, Ishida and Morisako 1985, Maier and Perego 1992, Araki and Hjelmstad 2000, Williamson and Hjelmstad 2001).

To the best of the authors' knowledge, the research on the dynamic collapse of structures under earthquake ground motions was initiated theoretically by Jennings and Husid (1968). They treated a single-degree-of-freedom (SDOF) system with an elastic-plastic spring and concluded that the P-delta effect lengthens the natural period of the structure and the model exhibits an exponentially large displacement approximately at the critical rotation where the coincidence of the resistance and the moment due to gravity force occurs. They also investigated the post-yield slope effect on collapse behaviors. Sun *et al.* (1973) discussed a similar condition by studying the free vibration of the system under an initial impact and derived the stability boundary in terms of initial velocity and displacement. Extension of the criterion for SDOF systems to multi-degree-of-freedom (MDOF) systems was challenged (Takizawa and Jennings 1980, Nakajima et al. 1990).

Many researches have also been conducted on the dynamic response of elastic-plastic structures by regarding the tangent stiffness as the key for characterizing instability. It is well recognized that if the tangent stiffness of an SDOF system becomes negative in a dynamic process, residual displacements increase. On the other hand, for MDOF systems, it was shown that a negative eigenvalue of the tangent stiffness matrix leads to either the accumulation of deformation in a particular mode (Uetani and Tagawa 1998) or to the localization of deformation (Maier and Perego 1992). Furthermore, Bernal (1998) indicated that a negative eigenvalue is a necessary condition of dynamic collapse. However, the existence of only a negative eigenvalue is not sufficient to induce dynamic collapse because the sign of the minimum eigenvalue can recover to be positive due to unloading. Hence, it was demonstrated that additional conditions considering unloading are necessary to predict dynamic collapse (Araki and Hjelmstad 2000).

In addition, the phenomenon of dynamic collapse of realistic frame models was investigated by several authors (Ger et al. 1993, Challa and Hall 1994, Hall 1998, Sivaselvan et al. 2009). In these studies, various effects, such as the spread of the plastic zone, nonlinear material behavior, and/or nonlinear geometric effects were incorporated in the numerical methods.

However, it does not seem that a reliable dynamic stability criterion has been proposed even for rather simple input. In this chapter, a simple closed-form dynamic stability criterion is presented for the first time for the double impulse as a simplified version of the near-fault ground motion. The closed-form expression is derived and used for demonstrating the fact that several patterns of unstable behaviors (collapse-process patterns) exist depending on the ratio of the input level of the double impulse to the structural strength and on the ratio of the negative post-yield stiffness to the initial elastic stiffness. The applicability of the presented method using the double impulse to actual recorded pulse-type ground motions is also investigated in the last of this chapter.

Historically the elastic-plastic earthquake responses were investigated for the steady-state response to sinusoidal input or the transient response to an extremely simple sinusoidal input in 1960–1970s (Caughey 1960a, b, Iwan 1961, 1965a, b). Those methods were applied to more complex problems including more complex structures and more complicated inputs. On the contrary, Kojima and Takewaki (2015a, b, c, 2016a) introduced a completely different approach and demonstrated that the peak elastic-plastic response (continuation of free-vibrations) can be derived by an energy balance approach without solving the equations of motion directly.

Dynamic stability or collapse criterion for elastic-plastic structures under near-fault ground motions is derived in this chapter in closed-form. Negative post-yield stiffness is treated in order to consider the P-delta effect. The double impulse is used as a substituted version of the fling-step near-fault ground motion. Since only free-vibration appears under such double impulse, the energy balance approach plays a critically important role in the derivation of the closed-form expression for a complicated elastic-plastic response of structures with the P-delta effect. It should be remarked that no iteration is needed in the derivation of the closed-form dynamic stability or collapse criterion on the critical elastic-plastic response. The closed-form expression enables the derivation of the fact that several patterns of unstable behaviors exist depending on the ratio of the input level of the double impulse to the structural strength and on the ratio of the negative post-yield stiffness to the initial elastic stiffness. The validity of the presented dynamic stability or collapse criterion is investigated by the numerical response analysis for structures under double impulses with stable or unstable (collapse or non-collapse) parameters. Furthermore, the reliability of the presented theory is investigated through the comparison with the response analysis to the corresponding one-cycle sinusoidal input as a representative version of the fling-step near-fault ground motion. The applicability of the proposed theory to actual recorded pulse-type ground motions is also discussed at the end of this chapter.

8.2 DOUBLE IMPULSE INPUT

8.2.1 Double impulse input

As explained in Chapters 2, 5, and 7 and in Kojima et al. (2015) and Kojima and Takewaki (2015a, b), the fling-step input (fault-parallel) of the near-fault ground motion can be represented effectively by a one-cycle sinusoidal wave (Mavroeidis and Papageorgiou 2003, Kalkan and Kunnath 2006) and the forward-directivity input (fault-normal) of the near-fault ground motion can be expressed by a series of three sinusoidal wavelets with different magnitudes (see Figure 1.2). It is explained in the field of seismology that the fling step is caused by the permanent displacement of the ground induced by the fault dislocation and the forward directivity effect is governed by the relation of the movement of the rupture front with the site. In this chapter, it is intended to simplify typical near-fault ground motions by a double impulse following the references (Kojima et al. 2015a, Kojima and Takewaki 2015a). This is because the double impulse in the form of shock has a simple characteristic and a straightforward expression of the response can be expected even for elastic-plastic responses based on an energy approach to free vibrations. Furthermore, the double impulse enables us to describe directly the critical timing of impulses (resonant frequency), which is not easy for the sinusoidal and other inputs without a repetitive procedure. While most of the previous methods (Caughey 1960a, b, Iwan 1961) employ the equivalent linearization of the structural model for the unchanged input (see Figure 8.1(a) including an equivalent linear stiffness), the method proposed in Kojima and Takewaki (2015a, b) and in this chapter transforms the input into the double impulse for the unchanged structural model (see Figure 8.1(b)). It should be noted that the negative post-yield slope cannot be dealt with by the equivalent linearization.

Following Kojima and Takewaki (2015a), consider a ground acceleration $\ddot{u}_g(t)$ as a double impulse, as shown in Figure 8.1(b), expressed by

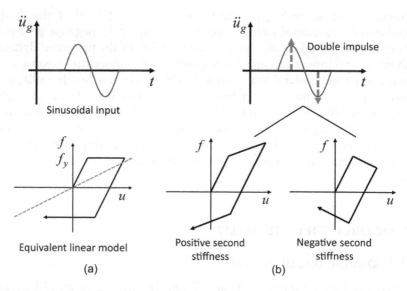

Figure 8.1 Feature of method of input transformation against method of structural model transformation: (a) previous method (equivalent linearization of structural model for unchanged input); (b) new method
(transformation of input into double impulse for unchanged structural
model).

$$\ddot{u}_g(t) = V\delta(t) - V\delta(t - t_0) \tag{8.1}$$

where V is the given initial velocity (also the second velocity with an opposite sign) and t_0 is the time interval between two impulses. The time derivative is denoted by an over-dot. Since the comparison with the corresponding
one-cycle sinusoidal wave from the viewpoint of response equivalence is
important, this will be conducted in Section 8.5.1.

The Fourier transform of the acceleration $\ddot{u}_g(t)$ of the double impulse can
be derived as

$$\ddot{U}_g(\omega) = \int_{-\infty}^{\infty} \{V\delta(t) - V\delta(t - t_0)\} e^{-i\omega t} dt = V(1 - e^{-i\omega t_0}) \tag{8.2}$$

8.2.2 Previous work on closed-form critical elastic–
perfectly plastic response of SDOF system
subjected to double impulse

In Chapter 2 and Kojima and Takewaki (2015a), a closed-form expression
of the critical response of an elastic–perfectly plastic (EPP) SDOF system
was derived for the double impulse. The critical response exhibiting the largest response under possible excitations plays a key role in the worst-case

analysis (Drenick 1970, Takewaki 2002, 2007, Moustafa et al. 2010, Takewaki et al. 2012). Since a similar classification of response cases is used in this chapter, the essence is shown in this section.

Consider an undamped elastic–perfectly plastic (EPP) SDOF system of mass m and stiffness k. The yield deformation and force are denoted by d_y and f_y. Let $\omega_1 = \sqrt{k/m}$, u, and f denote the undamped natural circular frequency, the mass displacement relative to the ground and the restoring force of the model, respectively. The plastic deformation just after the first impulse is expressed by u_{p1} and that just after the second impulse is denoted by u_{p2}.

The impulse changes the mass velocity by V instantaneously and the elastic-plastic response of the EPP SDOF system under the double impulse can be expressed by the continuation of free vibrations with different initial conditions. Let $u_{\max1}$ and $u_{\max2}$ denote the maximum deformations just after the first impulse and the second impulse, respectively, as shown in Figure 8.2. Those responses can be derived by an energy balance approach without solving the equation of motion directly. The kinetic energy given at the initial stage (the time of the first impulse) and at the time of the second impulse is transformed into the sum of the hysteretic energy and the maximum elastic strain energy corresponding to the yield deformation (see Section 1.3, 1.4 in Chapter 1). It was made clear that the critical timing, relative to the first impulse, of the second impulse corresponds to the state with a zero restoring force and only the kinetic energy exists at this stage as mechanical energies. By using this rule, the maximum deformation under the double impulse can be obtained in a simple manner.

The maximum response of the EPP SDOF system under the critical double impulse can be classified into the three cases depending on the yielding

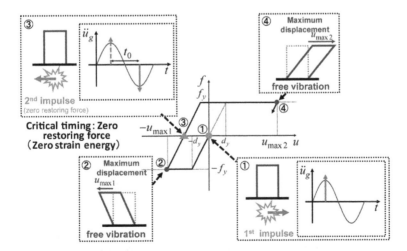

Figure 8.2 Schematic diagram of elastic–perfectly plastic SDOF model under critical double impulse.

stage (input level). Let $V_y(=\omega_1 d_y)$ denote the input velocity level of the double impulse at which the EPP SDOF system just attains the yield deformation just after the first impulse. This parameter also presents a strength parameter of the SDOF system. CASE 1 is the case of elastic response even after the second impulse and CASE 2 is the case of plastic deformation only after the second impulse. In addition, CASE 3 is the case of plastic deformation after the first impulse. Figure 8.3 shows the diagram for these three cases.

Figure 8.3(a) shows the maximum deformation just after the first impulse and that just after the second impulse, respectively, for the elastic case (CASE 1) during the whole stage. From the energy balance, $u_{\text{max}1}$ and $u_{\text{max}2}$ can be obtained as follows.

$$u_{\text{max}1}/d_y = V/V_y \tag{8.3}$$

$$u_{\text{max}2}/d_y = 2(V/V_y) \tag{8.4}$$

Using the similar energy balance, $u_{\text{max}1}$ and $u_{\text{max}2}$ for CASE 2 and CASE 3 (Figure 8.3(b), (c)) can be obtained simply as follows.

$$u_{\text{max}1}/d_y = V/V_y \quad (\text{CASE 2}) \tag{8.5}$$

$$u_{\text{max}2}/d_y = 0.5\left\{1+(2V/V_y)^2\right\} \quad (\text{CASE 2}) \tag{8.6}$$

$$u_{\text{max}1}/d_y = 0.5\left\{1+(V/V_y)^2\right\} \quad (\text{CASE 3}) \tag{8.7}$$

$$u_{\text{max}2}/d_y = 0.5(3+2V/V_y) \quad (\text{CASE 3}) \tag{8.8}$$

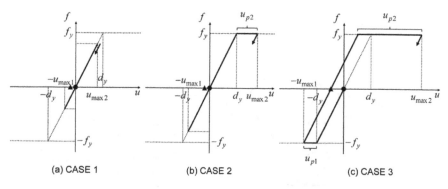

(a) CASE 1	(b) CASE 2	(c) CASE 3

Figure 8.3 Maximum deformation to double impulse based on energy approach: (a) CASE 1: Elastic response; (b) CASE 2: Plastic response only after second impulse; (c) CASE 3: Plastic response even after first impulse (●: first impulse, ▲: second impulse) (Kojima and Takewaki 2016a).

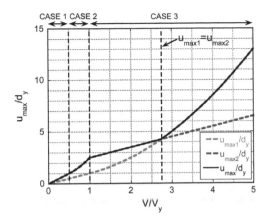

Figure 8.4 Maximum deformation to double impulse with respect to input level (Kojima and Takewaki 2015a).

Figure 8.4 illustrates the maximum deformation normalized by the yield deformation d_y with respect to input level.

8.3 MAXIMUM ELASTIC-PLASTIC DEFORMATION AND STABILITY LIMIT OF SDOF SYSTEM WITH NEGATIVE POST-YIELD STIFFNESS TO CRITICAL DOUBLE IMPULSE

Consider an elastic-plastic SDOF model with negative post-yield stiffness (EPN SDOF model) as shown in Figure. 8.5. The ratio of the post-yield stiffness to the initial elastic stiffness is denoted by $\alpha(<0)$. Other parameters are the same as those in the previous section. Let us introduce the notations shown in Figure 8.5. The plastic deformations after the first and second impulses are expressed by u_{p1} and u_{p2}, respectively, as in the previous section. In this chapter, the collapse of a structure (or stability limit) is characterized by the phenomenon that the restoring-force attains zero in the second stiffness range as shown in Figure 8.5.

8.3.1 Pattern 1: Stability limit after the second impulse without plastic deformation after the first impulse

The first collapse pattern is the case where the EPN SDOF model attains the stability limit after the second impulse without plastic deformation after the first impulse as shown in Figure 8.6. In order to derive the stability limit, the maximum deformation of the EPN SDOF model subjected to the critical double impulse is obtained by using the energy balance law.

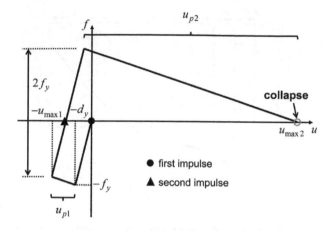

Figure 8.5 Definition of plastic deformations after first and second impulses and characterization of collapse of structure (Kojima and Takewaki 2016b).

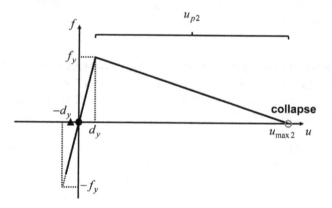

Figure 8.6 Restoring-force characteristic corresponding to Pattern I (stability limit after second impulse without plastic deformation after first impulse) (Kojima and Takewaki 2016b).

It can be proved that the critical timing of the second impulse to cause the maximum deformation after the second impulse is the time when the restoring-force becomes zero after the first impulse. At this timing, the velocity attains the maximum value V in the unloading process due to the energy conservation law and the velocity V is added just after the second impulse.

The plastic deformation of the EPN SDOF model just attaining the stability limit (zero restoring-force) after the second impulse can be obtained from Figure 8.6.

$$-\alpha k u_{p2} = k d_y \left(= f_y\right) \tag{8.9}$$

Eq. (8.9) leads to

$$u_{p2} = -(1/\alpha)d_y \qquad (8.10)$$

Then, the energy balance after the second impulse can be expressed by

$$m(2V)^2/2 = (f_yd_y/2) + f_yu_{p2} + (\alpha ku_{p2}^2/2) \qquad (8.11)$$

Substitution of Eq. (8.10) into Eq. (8.11) provides

$$m(2V)^2/2 = \{1-(1/\alpha)\}kd_y^2/2 \qquad (8.12)$$

Rearrangement of Eq. (8.12) with the use of the definition $\omega_1 d_y = V_y$ provides

$$(2V)^2 = \{1-(1/\alpha)\}\omega_1^2 d_y^2 = \{1-(1/\alpha)\}V_y^2 \qquad (8.13)$$

From Eq. (8.13), the input level of the double impulse at the stability limit can be expressed in terms of the post-yield stiffness ratio α as follows.

$$V/V_y = 0.5\sqrt{1-(1/\alpha)} \qquad (8.14)$$

In this pattern, $V/V_y \le 1.0$ has to be satisfied.
This stability limit corresponds to that by Sun et al. (1973).

8.3.2 Pattern 2: Stability limit after the second impulse with plastic deformation after the first impulse

The second collapse pattern is the case where the EPN SDOF model attains the stability limit after the second impulse with plastic deformation after the first impulse (see Figure 8.7). It can also be proved that the critical timing of the second impulse to cause the maximum deformation after the second impulse is the time when the restoring-force becomes zero after the first impulse. At this timing, the velocity attains the maximum value v_c due to the energy conservation law in the unloading process and the velocity V is added just after the second impulse.

From the energy balance law after the first impulse (see Figure 8.7), the plastic deformation u_{p1} can be obtained as

$$u_{p1} = -(1/\alpha)\left[1-\sqrt{1-\alpha\{1-(V/V_y)^2\}}\right]d_y \qquad (8.15)$$

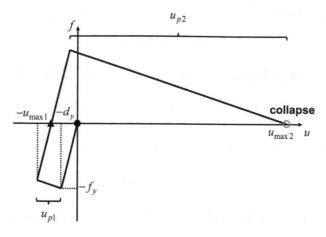

Figure 8.7 Restoring-force characteristic corresponding to Pattern 2 (stability limit after second impulse with plastic deformation after first impulse) (Kojima and Takewaki 2016b).

In addition, from the energy balance law in the unloading process just after attaining u_{p1} after the first impulse, the maximum velocity v_c during the unloading process can be expressed by

$$v_c = \sqrt{1 - \alpha \left\{ 1 - \left(V / V_y \right)^2 \right\}} V_y \qquad (8.16)$$

The relation of the plastic deformations u_{p1}, u_{p2} of the EPN SDOF model just attaining the stability limit (zero restoring-force) after the second impulse can be obtained from Figure 8.7.

$$-\alpha k u_{p2} = -f_y - \alpha k u_{p1} + 2f_y = f_y - \alpha k u_{p1} \qquad (8.17)$$

From Eq. (8.17), the plastic deformation u_{p2} can be expressed as

$$u_{p2} = -(1/\alpha) d_y + u_{p1} = -(1/\alpha) \left[2 - \sqrt{1 - \alpha \left\{ 1 - \left(V / V_y \right)^2 \right\}} \right] d_y \qquad (8.18)$$

Then, the energy balance after the second impulse (see Figure 8.7) can be expressed by

$$m(v_c + V)^2 / 2 = \left\{ k(d_y - \alpha u_{p1})^2 / 2 \right\} + \left(f_y - \alpha k u_{p1} \right) u_{p2} + \left(\alpha k u_{p2}^2 / 2 \right) \qquad (8.19)$$

Substitution of Eqs. (8.15), (8.16), and (8.18) into Eq. (8.19) provides

$$m\left[\sqrt{1-\alpha\left\{1-\left(V/V_y\right)^2\right\}}V_y+V\right]^2$$

$$=kd_y^{\,2}\left\{1-(1/\alpha)\right\}\left[2-\sqrt{1-\alpha\left\{1-\left(V/V_y\right)^2\right\}}\right]^2 \tag{8.20}$$

Rearrangement of Eq. (8.20) with the use of the definition $\omega_1 d_y = V_y$ and $k = \omega_1^2 m$ provides

$$\sqrt{1-\alpha\left\{1-\left(V/V_y\right)^2\right\}}V_y+V=\sqrt{-(1-\alpha)/\alpha}\left[2-\sqrt{1-\alpha\left\{1-\left(V/V_y\right)^2\right\}}\right]V_y \tag{8.21}$$

Eq. (8.21) can also be expressed as the following quadratic equation.

$$\left\{1-\alpha-\sqrt{-\alpha(1-\alpha)}\right\}\left(V/V_y\right)^2 - 4\sqrt{-(1-\alpha)/\alpha}\left(V/V_y\right)$$

$$-(1-\alpha)\left[(4/\alpha)+\left\{1+\sqrt{-(1-\alpha)/\alpha}\right\}^2\right]=0 \tag{8.22}$$

From Eq. (8.22), the input level of the double impulse at the stability limit can be expressed in terms of the post-yield stiffness ratio α as follows.

$$\frac{V}{V_y}=\frac{-2\sqrt{-(1-\alpha)/\alpha}\pm\sqrt{8\alpha^2-10\alpha+(2/\alpha)-2(4\alpha+1)(1-\alpha)\sqrt{-(1-\alpha)/\alpha}}}{2\left\{\alpha-1+\sqrt{-\alpha(1-\alpha)}\right\}} \tag{8.23}$$

In this pattern, $V/V_y \geq 1.0$ has to be satisfied. Therefore, one of two expressions in Eq. (8.23) is taken and reduced to the following form.

$$\frac{V}{V_y}=\frac{-2\sqrt{-(1-\alpha)/\alpha}-\sqrt{8\alpha^2-10\alpha+(2/\alpha)-2(4\alpha+1)(1-\alpha)\sqrt{-(1-\alpha)/\alpha}}}{2\left\{\alpha-1+\sqrt{-\alpha(1-\alpha)}\right\}} \tag{8.24}$$

8.3.3 Pattern 3: Stability limit after the second impulse with closed-loop in restoring-force characteristic

The third collapse pattern is derived here. From Figure 8.8, the energy balance during free vibration after attaining the maximum deformation after the second impulse can be expressed as

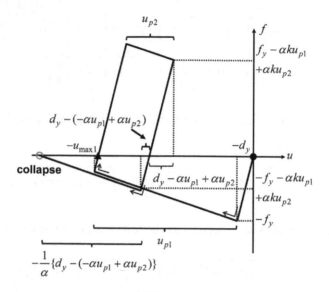

Figure 8.8 Restoring-force characteristic corresponding to Pattern 3 (stability limit after second impulse with closed-loop in restoring-force characteristic) (Kojima and Takewaki 2016b).

$$\frac{1}{2}k\left(d_y - \alpha u_{p1} + \alpha u_{p2}\right)^2 = \frac{1}{2}k\left\{d_y - \left(-\alpha u_{p1} + \alpha u_{p2}\right)\right\}^2$$
$$-\frac{1}{2\alpha}k\left\{d_y - \left(-\alpha u_{p1} + \alpha u_{p2}\right)\right\}^2 \qquad (8.25)$$

The parameters v_c and u_{p1} can be obtained from Eqs. (8.16) and (8.15), respectively, and u_{p2} can be derived from the energy balance law after the second impulse.

$$u_{p2} = \frac{1}{\alpha}\left[\left(\frac{v_c}{V_y} - 2\right) + \sqrt{\left(\frac{v_c}{V_y} - 2\right)^2 + \alpha\left\{\left(\frac{V}{V_y}\right)^2 + \left(2\frac{V}{V_y} + 4\right)\frac{v_c}{V_y} - 4\right\}}\right] \qquad (8.26)$$

With the notation $-\alpha u_{p1} + \alpha u_{p2} = \lambda d_y$, Eq. (8.25) provides

$$\frac{1}{2}k\left(d_y + \lambda d_y\right)^2 = \frac{1}{2}k\left(d_y - \lambda d_y\right)^2 - \frac{1}{2\alpha}k\left(d_y - \lambda d_y\right)^2 \qquad (8.27)$$

Division of both sides of Eq. (8.27) by $k/2$ and rearrangement of the resulting equation lead to

$$\left(1 + 2\lambda + \lambda^2\right)d_y^2 = \left(1 - 2\lambda + \lambda^2\right)d_y^2 - \frac{1}{\alpha}\left(1 - 2\lambda + \lambda^2\right)d_y^2 \qquad (8.28)$$

Division again of both sides of Eq. (8.28) by d_y^2 and rearrangement of the resulting equation yield

$$\lambda^2 - 2(1 - 2\alpha)\lambda + 1 = 0 \qquad (8.29)$$

The solution of Eq. (8.29) can be obtained as

$$\lambda = (1 - 2\alpha) \pm 2\sqrt{\alpha(\alpha - 1)} \qquad (8.30)$$

It should be noted that since $\lambda = (1 - 2\alpha) + 2\sqrt{\alpha(\alpha - 1)}$ provides only complex numbers, this case is eliminated. Then, the solution is reduced to

$$\lambda = (1 - 2\alpha) - 2\sqrt{\alpha(\alpha - 1)} \qquad (8.31)$$

Using Eqs. (8.15), (8.16), (8.26), (8.31), and $-\alpha u_{p1} + \alpha u_{p2} = \lambda d_y$, the following relation is derived.

$$\frac{\alpha}{d_y}(u_{p2} - u_{p1}) = -1 + \sqrt{\left(\frac{v_c}{V_y} - 2\right)^2 + \alpha\left\{\left(\frac{V}{V_y}\right)^2 + 2\left(\frac{V}{V_y} + 2\right)\frac{v_c}{V_y} - 4\right\}} = \lambda \quad (8.32)$$

After some rearrangement, Eq. (8.32) provides

$$\left(\frac{v_c}{V_y} - 2\right)^2 + \alpha\left\{\left(\frac{V}{V_y}\right)^2 + 2\left(\frac{V}{V_y} + 2\right)\frac{v_c}{V_y} - 4\right\} = (\lambda + 1)^2 \qquad (8.33)$$

Eqs. (8.16) and (8.33) lead to

$$2\alpha\left(\frac{V}{V_y}\right)^2 - 5\alpha + 5 - (\lambda + 1)^2 = -2\left(\alpha\frac{V}{V_y} + 2\alpha - 2\right)\frac{v_c}{V_y} \qquad (8.34)$$

After Eq. (8.34) is squared, the following fourth-order equation is obtained.

$$4\alpha^2\left(\frac{V}{V_y}\right)^4 + 4\alpha\left\{-5\alpha + 5 - (\lambda + 1)^2\right\}\left(\frac{V}{V_y}\right)^2$$
$$+ \left\{-5\alpha + 5 - (\lambda + 1)^2\right\}^2 = 4\left(\alpha\frac{V}{V_y} + 2\alpha - 2\right)^2\left(\frac{v_c}{V_y}\right)^2 \qquad (8.35)$$

Substitution of Eq. (8.16) into Eq. (8.35) yields

$$4\alpha^2(1-\alpha)\left(\frac{V}{V_y}\right)^4 + 16\alpha^2(1-\alpha)\left(\frac{V}{V_y}\right)^3$$

$$+4\alpha\left[\left\{-5\alpha+5-(\lambda+1)^2\right\}-\alpha(1-\alpha)-4(1-\alpha)^2\right]\left(\frac{V}{V_y}\right)^2$$

$$+16\alpha(1-\alpha)^2\left(\frac{V}{V_y}\right)+\left\{-5\alpha+5-(\lambda+1)^2\right\}^2-16(1-\alpha)^3=0$$

(8.36)

Further substitution of Eq. (8.31) into Eq. (8.36) and rearrangement of the resulting equation lead to

$$4\alpha^2\left(\frac{V}{V_y}\right)^4 + 16\alpha^2\left(\frac{V}{V_y}\right)^3$$

$$-4\alpha\left\{3-11\alpha-8\sqrt{-\alpha(1-\alpha)}\right\}\left(\frac{V}{V_y}\right)^2+16\alpha(1-\alpha)\left(\frac{V}{V_y}\right)$$

$$+(1-\alpha)\left\{128\alpha^2-32\alpha-15+16(8\alpha+1)\sqrt{-\alpha(1-\alpha)}\right\}=0$$

(8.37)

Eq. (8.37) provides the stability limit in this case, which depends on the parameter α.

8.3.4 Additional Pattern I: Limit after the first impulse

Another possible collapse pattern is the case where the EPN SDOF model attains the stability limit after the first impulse. Although the stability limit in this pattern is slightly larger than that for the above pattern 3, its limit is explained here for disclosing the overall property of the stability limit.

The plastic deformation u_{p1} of the EPN SDOF model just attaining the stability limit (zero restoring-force) after the first impulse can be obtained from Figure 8.9.

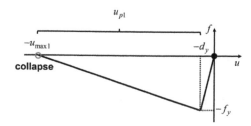

Figure 8.9 Additional Pattern I: Limit after first impulse (Kojima and Takewaki 2016b).

$$\alpha k u_{p1} = -k d_y \tag{8.38}$$

Eq. (8.38) leads to

$$u_{p1} = -(1/\alpha) d_y \tag{8.39}$$

The energy balance after the first impulse can be expressed by

$$m V^2 / 2 = (f_y d_y / 2) + f_y u_{p1} + (\alpha k u_{p1}^2 / 2) \tag{8.40}$$

Substitution of Eq. (8.39) into Eq. (8.40) provides

$$m V^2 / 2 = \{1 - (1/\alpha)\} k d_y^2 / 2 \tag{8.41}$$

Rearrangement of Eq. (8.41) with the use of $\omega_1 d_y = V_y$ provides

$$V^2 = \{1 - (1/\alpha)\} \omega_1^2 d_y^2 = \{1 - (1/\alpha)\} V_y^2 \tag{8.42}$$

From Eq. (8.42), the input level of the double impulse at the stability limit in this case can be expressed in terms of the post-yield stiffness ratio α as follows.

$$V/V_y = \sqrt{1 - (1/\alpha)} \tag{8.43}$$

In this pattern, $V/V_y \geq 1.0$ has to be satisfied.

8.3.5 Additional Pattern 2: Limit without plastic deformation after the second impulse

As in Section 8.3.4, since a further analysis of the overall behavior may be beneficial for accurate analysis of the stability limit, another classification analysis is made for the above-mentioned pattern 3. This classification is characterized by the condition whether the response after the second impulse shown in Figure 8.10 goes beyond the yield point.

From Figure 8.10, the energy balance after the second impulse in case of the elastic response after the second impulse can be expressed by

$$\frac{1}{2} m (v_c + V)^2 = \frac{1}{2} k \{u_e - (d_y + \alpha u_{p1})\}^2 \tag{8.44}$$

where u_e indicates the elastic deformation in the unloading process after attaining the maximum deformation u_{max1}. The maximum deformation after the second impulse can be derived as

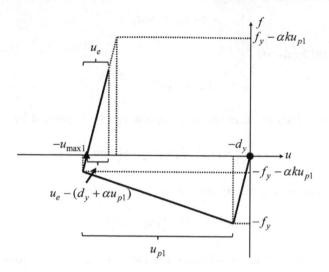

Figure 8.10 Additional Pattern 2: Limit without plastic deformation after second
impulse (Kojima and Takewaki 2016b).

$$u_{max2} = -u_{max1} + u_e = -d_y - u_{p1} + u_e \qquad (8.45)$$

The parameters v_c, u_{p1} were derived in Eqs. (8.16) and (8.15). From Eq.
(8.44), u_e can be obtained as

$$\frac{u_e}{d_y} = \left(\frac{v_c + V}{V_y}\right) + 1 + \alpha \frac{u_{p1}}{d_y} \qquad (8.46)$$

If u_e becomes larger than $2d_y$, the response goes into a plastic region after
the second impulse. Therefore, the boundary can be characterized by
$u_e = 2d_y$. Eq. (8.46) provides such a boundary as

$$\frac{u_e}{d_y} = \left(\frac{v_c + V}{V_y}\right) + 1 + \alpha \frac{u_{p1}}{d_y} = 2 \qquad (8.47)$$

Eq. (8.47) leads to the following quadratic equation.

$$(1 - 4\alpha)(V/V_y)^2 - 4(V/V_y) + 4\alpha = 0 \qquad (8.48)$$

The solution can be derived as

$$\frac{V}{V_y} = \frac{2 \pm 2\sqrt{4\alpha^2 - \alpha + 1}}{1 - 4\alpha} \qquad (8.49)$$

Since $V/V_y > 0$, the input velocity V/V_y corresponding to such boundary is expressed as

$$\frac{V}{V_y} = \frac{2 + 2\sqrt{4\alpha^2 - \alpha + 1}}{1 - 4\alpha} \tag{8.50}$$

It was confirmed that the EPN SDOF model goes into a plastic region after the second impulse in the input level $1.0 \le V/V_y \le \sqrt{1 - (1/\alpha)}$ from Additional Pattern 1, in the case of $\alpha > -1/3$. Therefore, the boundary given by Eq. (8.50) is used in the range of $\alpha < -1/3$.

Only the critical double impulse was dealt with in this chapter. It was confirmed that the stability limit for such critical timing plays a principal role in other noncritical cases (Homma et al. 2020). It should be reminded that the resonant case is not necessarily the worst case in a restricted situation.

8.4 RESULTS FOR NUMERICAL EXAMPLE

The dynamic stability (collapse) limit obtained in Section 8.3 is shown in Figure 8.11(a) where CASEs 1–3 indicate the response cases similar to those in Figure 8.3 for undamped EPP SDOF models. For better understanding, the corresponding collapse patterns are shown in Figure 8.11(a). The shaded areas in Figure 8.11(a) imply the unstable (collapse) area in the input level–post-yield stiffness relation. In order to investigate the accuracy of the presented limit, 15 points slightly smaller or larger than the limit curve for three post-yield stiffness ratios –0.1, – 1/3, – 0.6 were chosen. Those 15 points are indicated in Figure 8.11(b). The solid circles represent the stable (non-collapse) models, and the open circles present the unstable (collapse) models.

Figure 8.12 shows the restoring force-deformation relations for the above 15 models (three for the post-yield stiffness ratio –0.1, five for the post-yield stiffness ratio –1/3, and seven for the post-yield stiffness ratio –0.6). The colors correspond to the colors of circles in Figure 8.11(b) (for better understanding, see Kojima and Takewaki 2016b). It can be confirmed that the presented stability limit certainly divides the region into the stable one and the unstable one within a reasonable accuracy.

8.5 DISCUSSION

8.5.1 Applicability of critical double impulse timing to corresponding sinusoidal wave

Kojima and Takewaki (2015a) demonstrated that if the maximum value of the Fourier amplitude is selected as the key parameter (Section 1.2 in Chapter 1), the responses to the double impulse and the corresponding

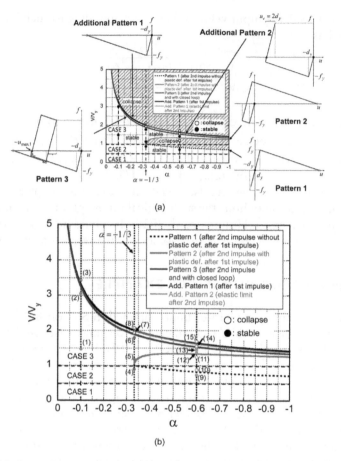

(a)

(b)

Figure 8.11 Several patterns of stability limit (patterns of collapse): (a) stability patterns; (b) fifteen points to be checked for stability (Kojima and Takewaki 2016b).

sinusoidal input exhibit a fairly good correspondence. In this section, it is investigated whether the critical timing derived from the double impulse is also an approximate critical timing of the sinusoidal input.

Let $t_0{}^c$ denote the critical timing of the double impulse and t_0 denote the general timing. The ratio of $t_0{}^c$ to the fundamental natural period $T_1(=2\pi/\omega_1)$ can be expressed as

$$\frac{t_0{}^c}{T_1} = \begin{cases} 0.5 & \text{for } V/V_y \le 1.0 \\ \frac{1}{2\pi}\arcsin\left(\frac{V_y}{V}\right) + \frac{1}{4\pi\sqrt{-\alpha}}\ln\left[\frac{1+\sqrt{-\alpha\left\{\left(V/V_y\right)^2-1\right\}}}{1-\sqrt{-\alpha\left\{\left(V/V_y\right)^2-1\right\}}}\right] + \frac{1}{4} & \text{for } V/V_y > 1.0 \end{cases} \tag{8.51}$$

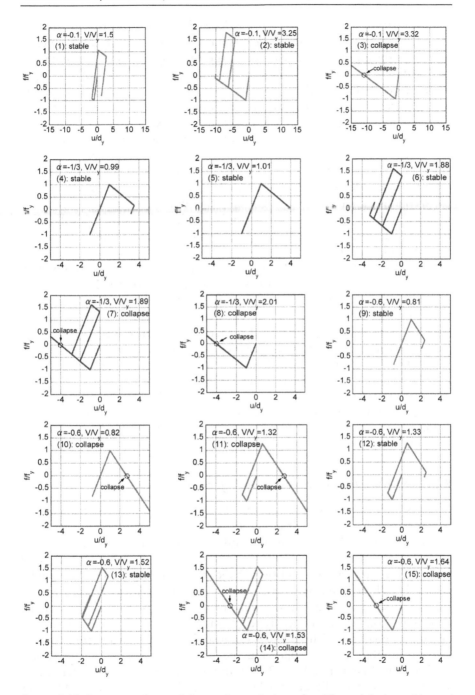

Figure 8.12 Restoring force-deformation relations for 15 models in stable and unstable regions (Kojima and Takewaki 2016b).

This relation is plotted in Figure 8.13(a) for several post-yield stiffness ratios. It can be seen that the critical timing is delayed due to plastic deformation as the input level increases, and the delaying rate is high as the post-yield stiffness ratio becomes smaller. The three vertical lines show the input levels corresponding to the additional pattern 1 where the collapse occurs after the first impulse. Since the second impulse cannot be input over this limit, Figure 8.13(a) represents this phenomenon exactly.

Figure 8.13(b) shows the maximum deformation of the model with $\alpha = -0.1$ under the corresponding sine wave with respect to $t_0/t_0{}^c$ where the sine wave has the circular frequency π/t_0 and the velocity amplitude is kept constant in each plot. The first peak for $V/V_y = 2.0$ corresponds to $u_{\mathrm{max}1}$ and the second peak indicates $u_{\mathrm{max}2}$. On the other hand, Figure 8.13(c) presents that of the model with $\alpha = -0.6$. It can be confirmed that the critical timing derived from the double impulse is also an approximate critical timing of the sinusoidal input.

It was also clarified from the numerical analysis using the corresponding one-cycle sinusoidal input that the stable models of Point (12) and (13) in Figure 8.11(b) are difficult to be produced. This fact may result from the fact that, since the deformation after the first impulse under the corresponding one-cycle sinusoidal input is smaller than the response under the double impulse, the energy absorption after the first impulse is small under the corresponding one-cycle sinusoidal input and the response after the second impulse under the corresponding one-cycle sinusoidal input goes easily into an unstable region. In such case ($\alpha < -1/3$), the model of Point (9) represents the model that is subjected to the double impulse with a slightly smaller input velocity than the stability limit. A more detailed examination should be made in the future.

8.5.2 Applicability to recorded ground motions

It seems important to investigate the applicability of the present theory to actual recorded pulse-type ground motions.

Consider the Rinaldi station fault-normal component during the Northridge earthquake in 1994 as a representative pulse-type ground motion. Since the ground motion is fixed, the structural model parameters should be selected appropriately, i.e., the parameter ω_1 or d_y in $V_y = \omega_1 d_y$ should be selected in an appropriate manner. Figure 8.14 illustrates the modeling of the part of the recorded ground motion acceleration into a one-cycle sinusoidal input. This one-cycle sinusoidal wave is further transformed into the double impulse following the method shown in the references (Kojima and Takewaki 2015a, 2016). In the transformation, the maximum values of Fourier amplitude have been coincided (see Section 1.2 in Chapter 1) and the interval of the double impulse is half the sinusoidal wave period. Since the initial velocity V (=1.636 [m/sec]) of the corresponding double impulse is determined in Figure 8.14, V_y is selected here. The corresponding

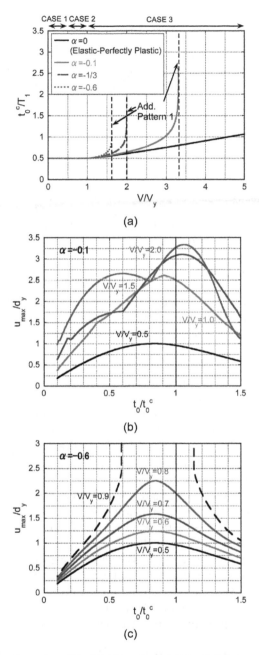

(a)

(b)

(c)

Figure 8.13 Critical timing of double impulse with respect to input level and correspondence of critical timing between double impulse and corresponding one-cycle sine wave: (a) critical timing of double impulse
with respect to input level; (b) maximum deformation of model with
$\alpha = -0.1$ under sine wave with respect to $t_0/t_0{}^c$; (c) maximum deformation of model with $\alpha = -0.6$ (Kojima and Takewaki 2016b).

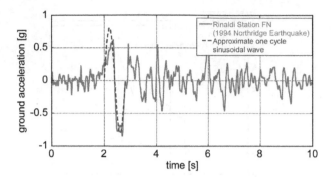

Figure 8.14 Modeling of part of pulse-type recorded ground motion into corresponding one-cycle sinusoidal input (Rinaldi station fault-normal component during the Northridge earthquake in 1994) (Kojima and Takewaki 2016b).

time interval t_0 is 0.4 [sec]. Because ω_1 is closely related to the resonance condition, d_y is selected principally. This procedure is similar to the well-known elastic-plastic response spectrum developed extensively in 1960–1970.

Figure 8.15(a) shows the maximum deformation with respect to V/V_y under the Rinaldi station fault-normal component and the corresponding double impulse. The solid line was drawn by using the method for response estimation shown in Appendix 1. On the other hand, the dotted line was obtained from the time-history response analysis for many models with different values of V_y. It can be found that $V/V_y = 0.8$ is the approximate limit. From the detailed investigation, $V/V_y = 0.78$ and $V/V_y = 0.79$ are selected for candidates to be investigated. These two models correspond approximately to Point (9) and (10) in Figure 8.11(b). Figure 8.15(b) demonstrates the restoring force-deformation relation for the stable case ($V/V_y = 0.78$) and the unstable case ($V/V_y = 0.79$) under the Rinaldi station fault-normal component. In addition, Figure 8.15(c) presents the deformation time-history for the stable case ($V/V_y = 0.78$) and the unstable case ($V/V_y = 0.79$) under the Rinaldi station fault-normal component. On the other hand, Figure 8.15 (d) shows the corresponding restoring-force time-history for the stable case ($V/V_y = 0.78$) and the unstable case ($V/V_y = 0.79$). It can be confirmed that the presented stability limit using the double impulse is fairly accurate.

Although V_y was varied in this section for a fixed set of V and t_0, it can also be understood that the result corresponds to the case where V_y is fixed and a set of V and t_0 is varied.

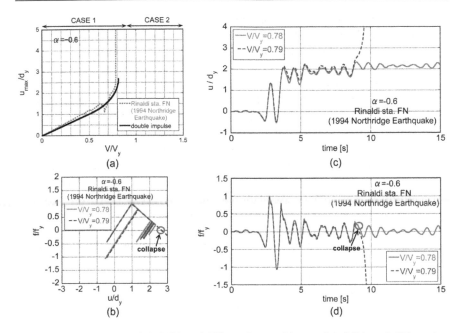

Figure 8.15 Stable model (V/V_y = 0.78) and unstable model (V/V_y = 0.79) under Rinaldi station fault-normal component: (a) maximum deformation with respect to V/V_y under Rinaldi station fault-normal component and corresponding double impulse; (b) restoring force-deformation relation for stable and unstable cases; (c) deformation time-history for stable and unstable cases; (d) restoring-force time-history for stable and unstable cases (Kojima and Takewaki 2016b).

8.6 SUMMARIES

A dynamic stability (or collapse) criterion for elastic-plastic structures under double impulse was derived in closed-form. The results may be summarized as follows:

1. The closed-form expression for the maximum elastic-plastic response of a bilinear hysteretic SDOF model under the critical double impulse (Kojima and Takewaki 2016c and Appendix 2) was extended to a dynamic stability problem of elastic-plastic structures with negative post-yield stiffness. A negative post-yield stiffness was treated to consider the P-delta effect. The double impulse was used as a substitute for the fling-step near-fault ground motion.

2. Since only free-vibration occurs under the double impulse, the energy balance approach plays a critical role in the derivation of the

closed-form expression for the maximum elastic-plastic response of the SDOF model with the P-delta effect. No iteration is required in the derivation of the closed-form dynamic stability criterion.

3. The closed-form expression made it clear that several patterns of unstable behaviors (collapse-process patterns) exist depending on the ratio of the input level of the double impulse to the structural strength and on the ratio of the negative post-yield stiffness to the initial elastic stiffness. The first pattern is the case where the structure attains the stability limit after the second impulse without plastic deformation after the first impulse. The second pattern is the case where the structure attains the stability limit after the second impulse with plastic deformation also after the first impulse. The third pattern is the case where the structure attains the stability limit after the second impulse with closed-loop in restoring-force characteristic (the final movement direction is the same as the first movement direction).

4. The validity of the criterion was investigated by the numerical response analysis for structures under double impulses with stable or unstable parameters. It was confirmed that the proposed criterion has reasonable accuracy.

5. The reliability of the theory was tested through the comparison with the response analysis to the corresponding one-cycle sinusoidal input as a representative of the fling-step near-fault ground motion.

6. It was demonstrated that the proposed criterion using the double impulse is applicable to actual recorded pulse-type ground motions within reasonable accuracy.

The theory for a damped system can be developed by using a similar energy balance law with the help of the method explained in Chapter 5 (Saotome et al. 2019). The present theory may also be applicable to an MDOF structure once the MDOF structure is transformed into the corresponding SDOF system using the pushover analysis as an example. Some readers interested in the general criterion for nonresonant case should access to the reference (Homma et al. 2020). It is remarkable that the resonant case is not necessarily the worst case in a restricted situation.

REFERENCES

Adam, C., and Jager, C. (2012). Dynamic instabilities of simple inelastic structures subjected to earthquake excitation, in *Advances Dynamics and Model-based Control of Structures and Machines*, Irschik, H., Krommer, M., and Belyaev, A. K. (eds.), Springer.

Araki, Y., and Hjelmstad, K. D. (2000). Criteria for assessing dynamic collapse of elastoplastic structural systems, *Earthquake Engng Struct. Dyn.*, 29: 1177–1198.

Bernal, D. (1987). Amplification factors for inelastic dynamic P-Δ effects in earthquake analysis, *Earthquake Engng Struct. Dyn.*, 15: 635–651.

Bernal, D. (1998). Instability of buildings during seismic response. *Eng. Struct.*, 20(4-6), 496–502.

Caughey, T. K. (1960a). Sinusoidal excitation of a system with bilinear hysteresis, *J. Appl. Mech.*, 27(4), 640–643.

Caughey, T. K. (1960b). Random excitation of a system with bilinear hysteresis, *J. Appl. Mech.*, 27(4), 649–652.

Challa, V. R. M., and Hall, J. F. (1994). Earthquake collapse analysis of steel frames. *Earthquake Engng Struct. Dyn.*, 23, 1199–1218.

Drenick, R. F. (1970). Model-free design of aseismic structures, *J. Eng. Mech. Div.*, ASCE, 96(EM4), 483–493.

Ger, J.-F., Cheng, Y., and Lu, L.-W. (1993). Collapse behavior of Pino Suarez building during 1985 Mexico City earthquake, *J. Struct. Eng.*, ASCE, 119(3), 852–870.

Hall, J. F. (1998). Seismic response of steel frame buildings to near-source ground motions, *Earthquake Engng Struct. Dyn.*, 27, 1445–1464.

Herrmann, G. (ed.) (1965). Dynamic stability of structures, *Proc. of Int. Conf. held at Northwestern University*, 1965, Pergamon Press, Oxford.

Hjelmstad, K. D., and Williamson E. B. (1998). Dynamic stability of structural systems subjected to base excitation, *Eng. Struct.*, 20(4–6), 425–432.

Homma, S., Kojima, K., and Takewaki, I. (2020). General dynamic collapse criterion for elastic-plastic structures under double impulse as substitute of near-fault ground motion, *Frontiers in Built Environment*, 6: 84.

Ibarra, L. F., and Krawinkler, H. (2005). Global collapse of frame structures under seismic excitations, PEER Center Report 2005/06, Richmond.

Ishida, S., and Morisako, K. (1985). Collapse of SDOF system to harmonic excitation, *J. Eng. Mech.*, ASCE, 111(3): 431–448.

Iwan, W. D. (1961). *The dynamic response of bilinear hysteretic systems*, Ph.D. Thesis, California Institute of Technology, Pasadena.

Iwan, W. D. (1965a). The dynamic response of the one-degree-of-freedom bilinear hysteretic system, *Proc. of the Third World Conf. on Earthquake Eng.*, New Zealand.

Iwan, W. D. (1965b). The steady-state response of a two-degree-of-freedom bilinear hysteretic system, *J. Applied Mech.*, 32(1), 151–156.

Jarernprasert, S., Bazan, E., and Bielak, J. (2013). Seismic soil-structure interaction response of inelastic structures, *Soil Dyn. Earthquake Eng.*, 47, 132–143.

Jennings, P. C., and Husid, R. (1968). Collapse of yielding structures during earthquakes, *J. Eng. Mech.*, ASCE, 94(EM5), 1045–1065.

Kalkan, E., and Kunnath, S. K. (2006). Effects of fling step and forward directivity on seismic response of buildings, *Earthquake Spectra*, 22(2), 367–390.

Kojima, K., Fujita, K., and Takewaki, I. (2015). Critical double impulse input and bound of earthquake input energy to building structure, *Frontiers in Built Environment*, 1: 5.

Kojima, K., and Takewaki, I. (2015a). Critical earthquake response of elastic-plastic structures under near-fault ground motions (Part 1: Fling-step input), *Frontiers in Built Environment*, 1: 12.

Kojima, K., and Takewaki, I. (2015b). Critical earthquake response of elastic-plastic structures under near-fault ground motions (Part 2: Forward-directivity input), *Frontiers in Built Environment*, 1: 13.

Kojima, K., and Takewaki, I. (2015c). Critical input and response of elastic-plastic structures under long-duration earthquake ground motions, *Frontiers in Built Environment*, 1: 15.

Kojima, K., and Takewaki, I. (2016a). Closed-form critical earthquake response of elastic-plastic structures on compliant ground under near-fault ground motions, *Frontiers in Built Environment*, 2: 1.

Kojima, K., and Takewaki, I. (2016b). Closed-form dynamic stability criterion for elastic-plastic structures under near-fault ground motions, *Frontiers in Built Environment*, 2: 6.

Kojima, K., and Takewaki, I. (2016c). Closed-form critical earthquake response of elastic-plastic structures with bilinear hysteresis under near-fault ground motions, *J. of Structural and Construction Eng.* (AIJ), 81(726): 1209–1219 (in Japanese).

Maier, G., and Perego, U. (1992). Effects of softening in elastic-plastic structural dynamics, *Int. J. Numerical Methods in Eng.*, 34, 319–347.

Mavroeidis, G. P., and Papageorgiou, A. S. (2003). A mathematical representation of near-fault ground motions, *Bull. Seism. Soc. Am.*, 93(3): 1099–1131.

Moustafa, A., Ueno, K., and Takewaki, I. (2010). Critical earthquake loads for SDOF inelastic structures considering evolution of seismic waves, *Earthquakes and Structures*; 1(2): 147–162.

Nakajima, A., Abe, H., and Kuranishi S. (1990). Effect of multiple collapse modes on dynamic failure of structures with structural instability. *Struct. Eng./ Earthquake Eng., JSCE*, 7(1): 1s–11s.

Saotome, Y., Kojima, K., and Takewaki, I. (2019). Collapse-limit input level of critical double impulse for damped bilinear hysteretic SDOF system with negative post-yield stiffness, *Frontiers in Built Environment*, 5: 106.

Sivaselvan, M. V., Lavan, O., Dargush, G. F., Kurino, H., Hyodo, Y., Fukuda, R., Sato, K., Apostolakis, G., and Reinhorn, A. M. (2009). Numerical collapse simulation of large-scale structural systems using an optimization-based algorithm, *Earthquake Engng Struct. Dyn.*, 38: 655–677.

Sun, C.-K., Berg, G. V., and Hanson R. D. (1973). Gravity effect on single-degree inelastic system, *J. Eng. Mech. Div.*, ASCE, 99(EM1): 183–200.

Takewaki, I. (2002). Robust building stiffness design for variable critical excitations, *J. Struct. Eng.*, ASCE, 128(12): 1565–1574.

Takewaki, I. (2007). *Critical excitation methods in earthquake engineering*, Elsevier, Amsterdam, Second edition in 2013, London.

Takizawa, H., and Jennings, P. C. (1980). Collapse of a model for ductile reinforced concrete frames under extreme earthquake motions, *Earthquake Engng Struct. Dyn.*, 8: 117–144.

Tanabashi, R., Nakamura, T., and Ishida, S. (1973). Gravity effect on the catastrophic dynamic response of strain-hardening multi-story frames, *Proc. of 5th World Conf. on Earthquake Eng.*, Rome, Italy, Vol. 2, 2140–2149.

Uetani, K., and Tagawa, H. (1998). Criteria for suppression of deformation concentration of building frames under severe earthquakes, *Eng. Struct.*, 20(4–6): 372–383.

Williamson, E. B., and Hjelmstad, K. D. (2001). Nonlinear dynamics of a harmonically-excited inelastic inverted pendulum, *J. Eng. Mech.*, ASCE, 127(1): 52–57.

APPENDIX 1: MAXIMUM ELASTIC-PLASTIC DEFORMATION OF SDOF MODEL WITH NEGATIVE POST-YIELD STIFFNESS TO DOUBLE IMPULSE

The maximum elastic-plastic deformation of an SDOF model with negative post-yield stiffness can be obtained based on the energy balance approach.

[CASE 1] Since the overall response is elastic as shown in Figure 8.A1(a), the maximum deformation was obtained in Kojima and Takewaki (2015a) and in Chapter 2 in this book.

[CASE 2] The maximum deformation u_{max} occurs after the second impulse ($u_{max} = u_{max2}$) as shown in Figure 8.A1(b). The maximum plastic deformation u_{p2} after the second impulse can be derived from the energy balance after the second impulse.

$$m(2V)^2/2 = f_y d_y/2 + f_y u_{p2} + \alpha k u_{p2}^2/2 \qquad (8.A1)$$

where $u_{max2} = d_y + u_{p2}$.

[CASE 3] The maximum plastic deformation u_{p1} after the first impulse can be obtained from the following energy balance after the first impulse (see Figure 8.A1(c)).

$$mV^2/2 = f_y d_y/2 + f_y u_{p1} + \alpha k u_{p1}^2/2 \qquad (8.A2)$$

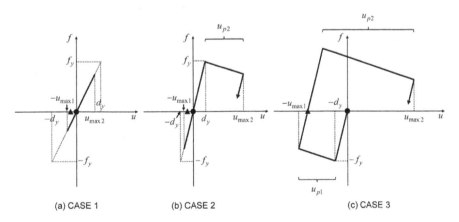

(a) CASE 1   (b) CASE 2   (c) CASE 3

Figure 8.A1   Maximum elastic-plastic deformation of SDOF model with negative post-yield stiffness to double impulse: (a) CASE 1: Elastic response; (b) CASE 2: Plastic response after the second impulse; (c) CASE 3: Plastic response after the first impulse (●: first impulse, ▲: second impulse) (Kojima and Takewaki 2016b).

where $u_{\text{max1}} = d_y + u_{p1}$. On the other hand, the maximum plastic deformation u_{p2} after the second impulse can be derived from the following energy balance after the second impulse (see Figure 8.A1(c)).

$$m\left(v_c + V\right)^2 / 2 = k\left(d_y - \alpha u_{p1}\right)^2 / 2 + \left(f_y - \alpha k u_{p1}\right)u_{p2} + \alpha k u_{p2}^2 / 2 \quad (8.\text{A}3)$$

where $u_{\text{max2}} = -u_{\text{max1}} + 2d_y + u_{p2}$ and v_c is the velocity at the zero restoring-force state. The velocity v_c can be obtained from the following energy balance after the starting point of the unloading process.

$$m v_c^2 / 2 = k\left(d_y + \alpha u_{p1}\right)^2 / 2 \quad (8.\text{A}4)$$

The maximum deformation u_{max} can be obtained as the larger value among u_{max1} derived from Eq. (8.A2) and u_{max2} derived from Eq. (8.A3).

For reference, the closed-form expression for the maximum elastic-plastic response of a bilinear hysteretic SDOF model with positive post-yield stiffness is shown in Appendix 2.

APPENDIX 2: MAXIMUM ELASTIC-PLASTIC DEFORMATION OF SDOF MODEL WITH POSITIVE POST-YIELD STIFFNESS TO DOUBLE IMPULSE

In this chapter, a bilinear hysteretic SDOF model with negative post-yield stiffness was treated. It may be useful to present the closed-form expression for the maximum elastic-plastic response of a bilinear hysteretic SDOF model with positive post-yield stiffness (Kojima and Takewaki 2016c).

Since the models with positive and negative post-yield stiffnesses can be treated in the same manner, the energy balance equations shown in Appendix 1 in this chapter can be used. Only the final expressions for the maximum elastic-plastic response are shown below.

[CASE 1] Since the overall response is elastic, the maximum deformation was obtained in Kojima and Takewaki (2015a) and in Chapter 2 in this book. The expression is shown below together with those for CASEs 2 and 3.

[CASE 2 & CASE 3]

$$\frac{u_{\text{max1}}}{d_y} = \begin{cases} V/V_y & \text{for} \quad 0 \le V/V_y < 1.0 \quad \text{(CASEs 1, 2)} \\ 1 + \left(u_{p1}/d_y\right) & \text{for} \quad 1.0 \le V/V_y \quad \text{(CASE 3)} \end{cases} \quad (8.\text{A}5)$$

$$\frac{u_{max2}}{d_y} = \begin{cases} 2V/V_y & \text{for } 0 \le V/V_y < 0.5 \,(\text{CASE 1}) \\[3mm] 1+\dfrac{1}{\alpha}\left[-1+\sqrt{1-\alpha\left\{1-\left(\dfrac{2V}{V_y}\right)^2\right\}}\,\right] & \text{for } 0.5 \le \dfrac{V}{V_y} < 1.0 \,(\text{CASE 2}) \\[3mm] -(u_{max1}/d_y)+2+(u_{p2}/d_y) & \text{for } 1.0 \le V/V_y \,(\text{CASE 3}) \end{cases} \quad (8.A6)$$

The plastic deformations u_{p1}/d_y, u_{p2}/d_y and the velocity v_c/V_y at the critical second impulse in CASE 3 can be expressed by

$$u_{p1}/d_y = (1/\alpha)\left[-1+\sqrt{1-\alpha\left\{1-(V/V_y)^2\right\}}\,\right] \quad (8.A7)$$

$$\frac{u_{p2}}{d_y} = \begin{cases} \dfrac{1}{\alpha}\left[\left(\dfrac{v_c}{V_y}-2\right)+\sqrt{\left(\dfrac{v_c}{V_y}-2\right)^2+\alpha\left\{\left(\dfrac{V}{V_y}\right)^2+\left(\dfrac{2V}{V_y}+4\right)\dfrac{v_c}{V_y}-4\right\}}\,\right] \\[2mm] \qquad\qquad\qquad\qquad \text{for } 1.0 \le V/V_y < \sqrt{1+3/\alpha} \\[3mm] \dfrac{u_{p1}}{d_y}-\dfrac{1}{\alpha}+\dfrac{1}{\sqrt{\alpha}}\dfrac{v_c+V}{V_y} \quad \text{for } \sqrt{1+\dfrac{3}{\alpha}} \le V/V_y \end{cases} \quad (8.A8)$$

$$\frac{v_c}{V_y} = \begin{cases} \sqrt{1-\alpha\left\{1-\left(\dfrac{V}{V_y}\right)^2\right\}} & \text{for } 1.0 \le \dfrac{V}{V_y} < \sqrt{1+\dfrac{3}{\alpha}} \\[3mm] \sqrt{\dfrac{1}{\alpha}\left(\alpha\dfrac{u_{p1}}{d_y}-1\right)^2+4\alpha\dfrac{u_{p1}}{d_y}} & \text{for } \sqrt{1+\dfrac{3}{\alpha}} \le V/V_y \end{cases} \quad (8.A9)$$

Chapter 9

Closed-form overturning limit of a rigid block as a SDOF model under near-fault ground motions

9.1 INTRODUCTION

The rocking response of rigid blocks under horizontal base input is of great concern in the evaluation of earthquake response of monuments, slender buildings, and furniture (or box on rack stores). In the seismic risk analysis of base-isolated high-rise buildings often constructed in Japan, this investigation is critical because most isolators do not have tensile resistance and overturning of such buildings should be prohibited at all. In the history of the earthquake structural engineering, the overturning rate of tombstones during an earthquake was often used in estimating the peak ground accelerations and velocities around the site.

The research on rocking response of rigid blocks under earthquake ground motions has been investigated extensively since the pioneering work by Milne (1885) and Housner (1963). It is well known that the rocking behavior of rigid blocks is strongly nonlinear and strong backgrounds from mathematical and physical viewpoints are inevitable for exact analysis. Yim et al. (1980) conducted an extensive investigation for many recorded ground motions based on the work by Housner (1963). Ishiyama (1982) studied various types of the nonlinearity of the overturning response of a rigid block in detail. After these works, many investigations have been conducted so far (Priestley et al. 1978, Spanos and Koh 1984, Hogan 1989, 1990, Shenton III and Jones 1991, Pompei et al. 1998, Andreaus and Casini 1999, Anooshehpoor et al. 1999, Zhang and Makris 2001, Prieto et al. 2004, Yilmaz et al. 2009, ElGawady et al. 2010, DeJong 2012, Dimitrakopoulos and DeJong 2012a, b). Recently, DeJong (2012) and Dimitrakopoulos and DeJong (2012a, b) investigated the rocking motion and overturning of a rigid block in detail and derived important results. Casapulla et al. (2010) and Casapulla and Maione (2016) considered the multiple sequences of impulses for the rocking response of a rigid block and derived the resonant response. They introduced two types of multiple impulses, i.e., with gradually increasing intervals for resonance and with equal intervals. Furthermore, Makris and Kampas (2016) investigated the scale effect of blocks on the overturning limit level of sinusoidal inputs and earthquake ground motions.

The scale effect is very important in the reliable design of tall and mega structures.

It is also important to focus on the history of nonlinear response analysis of structures. In an early stage of structural dynamics, the resources of computers are quite limited and the elastic-plastic earthquake responses were investigated primarily for the steady-state response to harmonic input or the transient response to an extremely simple sinusoidal input in 1960–1970s (Caughey 1960a, b, Iwan 1961, 1965a, b). After the development of methods for sophisticated mathematical description (e.g., finite element methods) of those responses, such simple techniques have been applied to more complex problems. On the other hand, although applicable only for simplified inputs, Kojima and Takewaki (2015a, b, c) and Kojima and Takewaki (2016a, 2016b) proposed recently a completely different approach using a double impulse in place of near-fault ground motions and demonstrated that the peak elastic-plastic response can be derived by using an energy balance approach without solving the equations of motion directly even for models with negative post-yield stiffness.

A closed-form limit value on the input level of the double impulse as a substituted version of the near-fault ground motion is derived for the overturning phenomenon of a rigid block in this chapter. The rocking vibration of the rigid block is formulated by using both the conservation law of angular momentum and the conservation law of mechanical energy. The initial rotational velocity after the first impulse and the rotational velocity after the impact are determined by the conservation law of angular momentum. The velocity change after the second impulse is also characterized by the conservation law of angular momentum. The maximum angles of rotation of the rigid block in both the clockwise and counterclockwise directions, which are needed for the computation of the overturning limit, are derived by the conservation law of mechanical energy. This enables us to avoid the computation of complicated nonlinear step-by-step time-history responses. The critical timing of the second impulse to the first impulse is characterized by the time of impact after the first impulse. It is clarified that the action of the second impulse just after the impact corresponds to the critical timing. The closed-form expression for the critical velocity amplitude limit of the double impulse enables the derivation of the important fact that its limit is proportional to the square root of size, i.e., the scale effect.

9.2 DOUBLE IMPULSE INPUT

In this chapter, it is intended to model a principal part of the near-fault ground motion into a one-cycle sinusoidal wave (Kalkan and Kunnath 2006) and then simplify such one-cycle sinusoidal wave into a double impulse following the references (Kojima et al. 2015, Kojima and Takewaki 2015a, c, 2016a, b, Taniguchi et al. 2016) as shown in Figure 9.1. This is

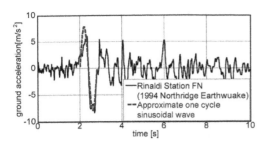
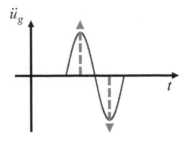

Figure 9.1 Modeling of principal part of near-fault ground motion into one-cycle sinusoidal wave and modeling of such one-cycle sinusoidal wave into double impulse (Nabeshima et al. 2016).

because the double impulse in the form of shock has a simple characteristic and a straightforward expression of the response can be expected even for nonlinear elastic responses based on an energy approach to free vibrations.

Following the reference (Kojima and Takewaki 2015a), consider a ground acceleration $\ddot{u}_g(t)$ as a double impulse, as shown in Figure 9.1, expressed by

$$\ddot{u}_g(t) = V\delta(t) - V\delta(t - t_0) \tag{9.1}$$

where V is the given initial velocity (also the second velocity with an opposite sign), $\delta(t)$ is the Dirac delta function, and t_0 is the time interval between two impulses. As in other chapters, the time derivative is denoted by an overdot. The comparison with the corresponding one-cycle sinusoidal wave is plotted in Figure 9.1. The corresponding velocity and displacement of such double impulse and sinusoidal wave can also be found in Kojima and Takewaki (2015a) and Chapter 1. It was confirmed that the double impulse is a good approximation of the corresponding sinusoidal wave even in the form of velocity and displacement. For the good correspondence in the response, the velocity amplitude of the double impulse has to be adjusted to the acceleration amplitude of the corresponding sine wave with the help of the equivalence of the maximum Fourier amplitude (see Section 1.2 in Chapter 1).

The Fourier transform of the acceleration $\ddot{u}_g(t)$ of the double impulse can be derived as

$$\ddot{U}_g(\omega) = \int_{-\infty}^{\infty} \{V\delta(t) - V\delta(t - t_0)\} e^{-i\omega t} dt = V(1 - e^{-i\omega t_0}) \tag{9.2}$$

As a comparative approach, an equivalent linearization method exists. While most of the previous methods (Caughey 1960a, b, Iwan 1961) employ the equivalent linearization of the structural model for the unchanged input (see Figure 9.2(a) including an equivalent linear stiffness), the method

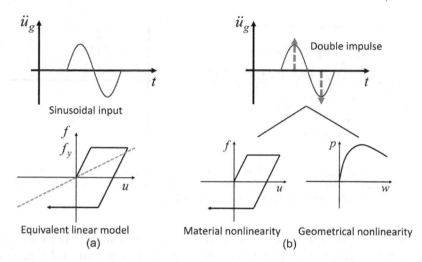

Figure 9.2 Input model transformation or structural model transformation: (a) previous method (equivalent linearization of structural model for unchanged input); (b) novel method (transformation of input into double impulse for unchanged nonlinear structural model).

presented in Kojima and Takewaki (2015a) and in this chapter transforms the input into the double impulse for the unchanged structural model (see Figure 9.2(b)). It should be noted that the negative second slope cannot be dealt with by the equivalent linearization.

9.3 MAXIMUM ROTATION OF RIGID BLOCK SUBJECTED TO CRITICAL DOUBLE IMPULSE

Consider the rocking response of a rigid block of mass m with width $2b$ and height $2h$ under a base horizontal acceleration input $\ddot{u}_g(t)$ as shown in Figure 9.3 (input of $-\ddot{u}_g(t)$, Eq. (9.1) with minus sign, does not cause any problem). The geometrical properties can be represented by two parameters, the length $R = \sqrt{b^2 + h^2}$ and angle α (= tan (b/h)), as shown in Figure 9.3. Let $I(=(4/3)mR^2)$ and g denote the mass moment of inertia around the edge of bottom right (also bottom left) and the acceleration of gravity, respectively. In the field of rocking response of a rigid block, it is a common understanding that this model can be substituted by a rigid bar, as shown in Figure 9.3, with the same mass moment of inertia supported by a nonlinear elastic rotational spring with rigid initial stiffness and negative nonlinear second slope. The moment-rotation relation of the nonlinear elastic rotational spring with rigid initial stiffness and the negative nonlinear second slope is shown in Figure 9.4. The schematic diagram of the response of the rigid

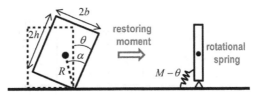

Figure 9.3 Modeling of rocking rigid block by rigid bar supported by nonlinear elastic rotational spring with rigid initial stiffness and negative second slope (Nabeshima et al. 2016).

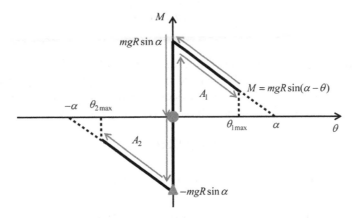

Figure 9.4 Moment-rotation relation for rocking response of rigid block and timing of double impulse (Nabeshima et al. 2016).

block to the critical double impulse (input of $-\ddot{u}_g(t)$, i.e., Eq. (9.1) with minus sign) is shown in Figure 9.5.

As in most investigations on rocking response of a rigid block, slipping of the block is ignored here for simplicity. Furthermore, a scenario that the overturning occurs after the action of the second impulse is employed in this chapter. This scenario seems valid because the input limit on the overturning corresponding to this scenario provides a lower limit in general. As for more detailed scenarios, see Ishiyama (1982) and Dimitrakopoulos and DeJong (2012b).

The critical acting timing of the second impulse is at the impact where the rotational velocity attains the maximum (see Kojima and Takewaki 2015a). Furthermore, it can be shown that critical timing is just after the impact because the rotational velocity is reduced greatly at the impact. If the second impulse is input just before the impact, its effect on the overturning is reduced greatly by the reduction of the velocity. This discussion is very important in the rocking response of a rigid block under the double impulse. More detailed verification of the critical timing of the second impulse is shown in Appendix 1.

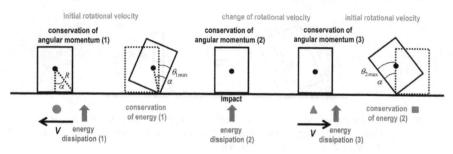

Figure 9.5 Rocking response of rigid block and governing law (conservation of angular momentum, conservation of energy, energy dissipation) (Nabeshima et al. 2016).

Let $\theta(t)$ denote the angle of rotation of the rigid block (clockwise direction is positive). Since the double impulse (input of $-\ddot{u}_g(t)$, i.e., Eq. (9.1) with minus sign) is set, the first rotation occurs in the clockwise direction. The equation of motion for this rigid block can be expressed by

$$I\ddot{\theta}(t) + mgR\sin\{-\alpha - \theta(t)\} = -m\ddot{u}_g(t)R\cos\{-\alpha - \theta(t)\} \quad \theta(t) < 0 \quad (9.3a)$$

$$I\ddot{\theta}(t) + mgR\sin\{\alpha - \theta(t)\} = -m\ddot{u}_g(t)R\cos\{\alpha - \theta(t)\} \quad \theta(t) > 0 \quad (9.3b)$$

The maximum angle of rotation after the first impulse is expressed by $\theta_{1\max}$ and that after the second impulse is denoted by $\theta_{2\max}$.

Since the rigid block stands on the floor at first, the exact description of the input of the first impulse and its effect on the initial response of the rigid block is very important for later analysis. The first conservation law of angular momentum (conservation of angular momentum (1) in Figure 9.5) can be expressed by

$$mVR\cos\alpha = I\dot{\theta}_1 \qquad (9.4)$$

From Eq. (9.4), the initial rotational velocity $\dot{\theta}_1$ after the first impulse can be obtained as

$$\dot{\theta}_1 = mVR\cos\alpha/I \qquad (9.5)$$

The first conservation law of mechanical energy (conservation of energy (1) in Figure 9.5) after substitution of Eq. (9.5) yields

$$(1/2)I\dot{\theta}_1^2 = (1/2)I(mVR\cos\alpha/I)^2 = A_1 \qquad (9.6)$$

A_1 is the area of near trapezoid corresponding to $\theta_{1\max}$ shown in Figure 9.4. It can be understood from Eq. (9.6) that once the velocity amplitude V of the double impulse is given, $\theta_{1\max}$ can be obtained.

The second conservation law of angular momentum (conservation of angular momentum (2) in Figure 9.5) at the impact ($\dot{\theta}_1$: rotational velocity just before the impact) can be given by

$$I\dot{\theta}_1 - 2mRb\dot{\theta}_1 \sin\alpha = I\dot{\theta}_2^{(1)} \tag{9.7}$$

This conservation law is applied at the pivot point (corner of the block). Since the rotation occurs around the bottom right before the impact, two terms appear on the left side of Eq. (9.7). Following the reference (Housner 1963), the rotational velocity $\dot{\theta}_2^{(1)}$ just after the impact may be expressed by

$$\dot{\theta}_2^{(1)} = \sqrt{r}\dot{\theta}_1 = \sqrt{r}\left(\frac{mVR\cos\alpha}{I}\right) \tag{9.8}$$

where r can be obtained from Eq. (9.7) as

$$r = \left\{1 - (2mRb\sin\alpha/I)\right\}^2 \tag{9.9}$$

This parameter r was introduced first by Housner (1963). When the boundary condition between the block and the base is necessary to consider, e.g., the surface material properties (ElGawady et al. 2010) or the rocking of tall buildings (limited contact area), another coefficient should be added on the parameter r.

Application of the conservation law of angular momentum (conservation of angular momentum (3) in Figure 9.5) just after the second impulse leads to the rotational velocity change $\dot{\theta}_2^{(2)}$ by the second impulse.

$$I\dot{\theta}_2^{(2)} = mVR\cos\alpha \tag{9.10}$$

In this case, the rotational velocity $\dot{\theta}_2$ just after the second impulse can be derived as

$$\dot{\theta}_2 = \dot{\theta}_2^{(1)} + \dot{\theta}_2^{(2)} \tag{9.11}$$

The rotation angle $\dot{\theta}_2^{(1)}$ is obtained based on the Housner's formulation (conservation law of angular momentum at the impact) and $\dot{\theta}_2^{(2)}$ is derived from the transformation of the horizontal impulse into rotation (the same treatment is made in Eq. (9.4)). Since these two phenomena occur in time sequence, a summation of the angles of rotation can be verified.

The second conservation law of mechanical energy (conservation of energy (2) in Figure 9.5) provides the following expression.

$$(1/2)I\dot{\theta}_2^2 = A_2 \tag{9.12}$$

A_2 is the area of a nearly trapezoid corresponding to $\theta_{2\max}$ shown in Figure 9.4.

It may be interesting to note the quantity of energy dissipation during the rocking response. The energy dissipation (energy dissipation (1) in Figure 9.5) can be expressed by

$$(1/2)mV^2 \rightarrow (1/2)I\dot{\theta}_1^2 = (1/2)mV^2 \times (3/4)\cos^2\alpha \qquad (9.13)$$

On the other hand, the energy dissipation (energy dissipation (2) in Figure 9.5) can be described by

$$(1/2)I\dot{\theta}_1^2 \rightarrow (1/2)I\dot{\theta}_2^{(1)2} = (1/2)I\dot{\theta}_1^2 \times r \qquad (9.14)$$

The critical timing t_0 can be obtained approximately (linear approximation: see the reference (Housner 1963) for more detail) by solving the equation of free-rocking motion from the first impulse to the second impulse.

$$t_0 = \frac{2}{p}\cosh^{-1}\left\{\frac{1}{1-(\theta_{1\max}/\alpha)}\right\} \qquad (9.15)$$

where $p^2 = mgR/I$. It should be reminded that $\theta_{1\max}$ was obtained in Eq. (9.6).

9.4 LIMIT INPUT LEVEL OF CRITICAL DOUBLE IMPULSE CHARACTERIZING OVERTURNING OF RIGID BLOCK

Consider the overturning phenomenon in the next step. The overturning of the rigid block can be characterized by the attainment of $\theta_{2\max}$ to $-\alpha$ as shown in Figure 9.6.

$$\theta_{2\max} = -\alpha \qquad (9.16)$$

In this case, the nearly trapezoidal shape corresponding to the area A_2 in Figure 9.6 is reduced to the nearly triangle shape. Eq. (9.12) can then be expressed as follows.

$$(1/2)I\dot{\theta}_2^2 = \int_0^\alpha mgR\sin(\alpha-\theta)d\theta \qquad (9.17)$$

Let us substitute Eqs.(9.8) and (9.10) into Eq. (9.11) and then the resulting expression into Eq. (9.17). Then, Eq. (9.17) can be reduced to

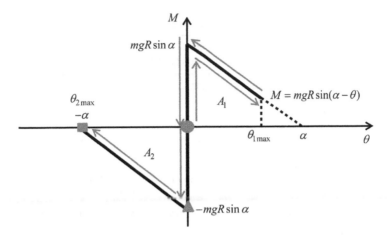

Figure 9.6 Moment-rotation relation for rocking response of rigid block and limit of overturning: solid circle, solid triangle, and solid rectangle indicate stages shown in Figure 9.5 (Nabeshima et al. 2016).

$$\frac{1}{2}\left(\frac{4}{3}mR^2\right)\left\{\frac{3hV}{4R^2}\left(1+\sqrt{r}\right)\right\}^2 = mgR\left(1-\cos\alpha\right) = mg\left(R-h\right) \quad (9.18)$$

The critical velocity amplitude (overturning limit velocity) of the double impulse can then be obtained as follows.

$$V_c = \frac{2R}{\left(1+\sqrt{r}\right)h}\sqrt{\frac{2\left(R-h\right)}{3}g} \quad (9.19)$$

It should be pointed out that the critical velocity amplitude of the double impulse is proportional to the square root of size. This finding is an important instruction for the structural design of monuments and tall buildings. In other words, as the monuments and tall buildings become larger, they become more stable and tough for overturning.

The critical timing t_0 can be evaluated as the time between the first impulse and the second impulse. This quantity can be obtained by solving the linearized equation of motion in the positive rotation range. When the critical condition is substituted and linearization is introduced in the evaluation of $\theta_{1\text{max}}$, Eq. (9.15) is reduced to:

$$t_0 = \frac{2}{p}\cosh^{-1}\left\{\frac{1+\sqrt{r}}{\sqrt{r+2\sqrt{r}}}\right\} \quad (9.20)$$

The flowchart for finding the critical velocity amplitude of the double impulse is shown in Figure 9.7.

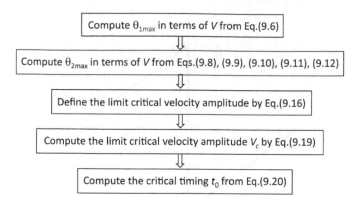

Figure 9.7 Flowchart for finding critical velocity amplitude of double impulse.

When the equivalent one-cycle sinusoidal wave (period $T_p=2t_0$) is treated, it can be shown that the velocity amplitude V_p of the equivalent one-cycle sinusoidal wave (the maximum Fourier amplitude is equivalent (Kojima and Takewaki 2015a and Section 1.2 in Chapter 1) is proportional to V of the double impulse ($V_p/V = 1.22218898...$). In this case, the acceleration amplitude A_p can be obtained from $A_p = \omega_p V_p/2$ where $\omega_p = 2\pi/T_p$.

9.5 NUMERICAL EXAMPLES AND DISCUSSION

In order to demonstrate the accuracy and reliability of the method explained in this chapter, numerical examples are presented. Consider three numerical examples of rectangular columns with width ($2b$) = 1, 2, 4m corresponding to the reference due to Makris and Kampas (2016). The column height is changed parametrically.

Figure 9.8 shows the limit velocity amplitude V_c of the critical double impulse with respect to the diagonal of the rigid block R for the above-mentioned three models computed by Eq. (9.19). The limit of velocity amplitude by the numerical simulation of time-history response is also plotted in

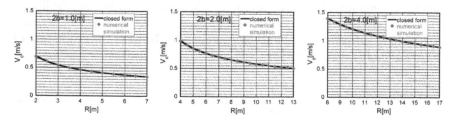

Figure 9.8 Limit velocity amplitude of critical double impulse with respect to R for $2b$ = 1, 2, 4[m] (closed-form expression and numerical simulation) (Nabeshima et al. 2016).

Figure 9.8. This numerical simulation limit was obtained by changing the velocity amplitude to the overturning together with the corresponding critical impulse timing which was derived by the present formulation, i.e., Eq. (9.20). The numerical integration was conducted by the fourth-order Runge-Kutta method (the time increment = 0.001s). The presented closed-form limit velocity amplitude of the critical double impulse coincides fairly well with the numerical simulation result and is quite reliable.

Figure 9.9 presents the critical timing given by Eq. (9.20) with respect to the diagonal of the rigid block R for the above-mentioned three models together with the plot by the numerical simulation for the nonlinear model. In the numerical simulation, the equation of free-rocking motion of the rigid block was solved numerically (numerically integrated), and the time interval between the first impulse and the second impulse was employed as the critical timing. It should be reminded that the closed-form critical timing is based on a linear approximation of the equation of motion as in the reference (Housner 1963), and a slight difference with the result by the numerical simulation for the nonlinear case can be seen in the range of smaller height. However, it appears that this difference is negligible.

It may be useful to compare the present overturning limit velocity with the previously proposed one. To do this, the comparison in the acceleration scale is necessary. Figure 9.10 shows the limit acceleration amplitude divided by the acceleration of gravity of the equivalent one-cycle sinusoidal input.

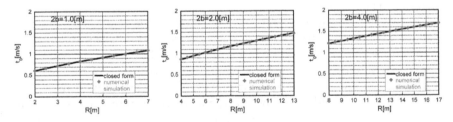

Figure 9.9 Critical timing with respect to R for 2b = 1, 2, 4[m] (closed-form expression for linearized model and numerical simulation for nonlinear model) (Nabeshima et al. 2016).

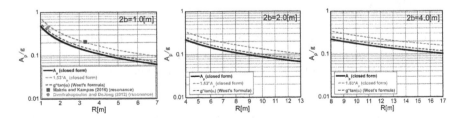

Figure 9.10 Critical acceleration amplitude ratio of equivalent one-cycle sinusoidal input to acceleration of gravity for 2b = 1, 2, 4[m] and comparison with other results (Nabeshima et al. 2016).

The amplitude is evaluated by the transformation mentioned above, i.e., $A_p = \omega_p V_p/2$, $V_p/V_c = 1.22218898...$, $\omega_p = 2\pi/T_p$, $T_p = 2t_0$ (Section 1.2 in Chapter 1). In addition, Figure 9.10 illustrates the magnified plot with factor 1.53. This factor was introduced from the viewpoint of the equivalence of response between the double impulse (also the maximum-Fourier-amplitude-equivalent one-cycle sinusoidal wave) and the rocking-response-equivalent one-cycle sinusoidal wave. This magnification will be discussed in the following (Figure 9.11). Other data from Makris and Kampas (2016), Dimitrakopoulos and DeJong (2012b) (coefficient of restitution = 0.8), and the West's formula (Milne 1885) were also plotted for comparison. The plots from Makris and Kampas (2016) and Dimitrakopoulos and DeJong (2012b) are limited because only the resonant one was picked up. It can be observed that the present magnified expression corresponds fairly well to other results for resonance.

Figure 9.11 presents several comparisons ($2b = 1, 2, 4$[m]) of time-history responses $\theta(t)/\alpha$ at the overturning limit between the double impulse and the corresponding equivalent one-cycle sinusoidal wave magnified by a certain coefficient (A_{sine}: amplitude of magnified sine wave). It was found that when the amplitude of the equivalent one-cycle sinusoidal wave is magnified by a coefficient about 1.53–1.54, both response amplitudes coincide fairly well. This phenomenon may result from the fact that while the resonance is guaranteed for the double impulse, that is not for the equivalent one-cycle sinusoidal wave. The difference of $\dot{\theta}_1$ just before the impact may be another factor of the discrepancy.

The two-step transformation of the magnitude of the equivalent sinusoidal waves consists of the introduction of the one-cycle sinusoidal wave equivalent to the double impulse in Section 9.4 and the magnification of the equivalent one-cycle sinusoidal wave in this section. The introduction of this two-step transformation comes from the viewpoint of the equivalence of the maximum Fourier amplitude and from the viewpoint of connection to the previous related works by the present authors. If the readers prefer the correction factor, they can multiply $V_p/V = 1.22218898...$ and $A_{sine}/A_p = 1.53 - 1.54$.

9.6 SUMMARIES

An explicit limit on the input velocity level of the double impulse as a representative of the principal part of a near-fault ground motion has been derived for the overturning of a rigid block. The results may be summarized as follows.

1. The rocking vibration of a rigid block was formulated by using the conservation law of angular momentum and the conservation law of mechanical energy. The conservation law of angular momentum was used in determining the initial rotational velocity just after the first

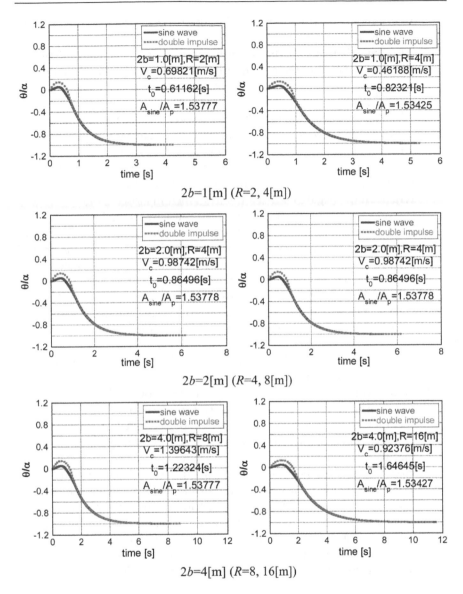

Figure 9.11 Comparison of time-history responses $\theta(t)/\alpha$ at overturning limit between double impulse and corresponding equivalent one-cycle sinusoidal wave magnified by coefficient about 1.53–1.54. (Nabeshima et al. 2016).

impulse and the rotational velocity change at the impact on the ground and the input of the second impulse. On the other hand, the conservation law of energy was used in obtaining the maximum rotational angles after the first impulse and that after the second impulse, which are needed for the computation of the overturning limit. This enabled us to avoid the computation of complicated nonlinear time-history responses.

2. The critical timing of the second impulse was characterized by the time of impact after the first impulse. It was clarified that the action of the second impulse just after the impact on ground corresponds to the critical timing.

3. The overturning of the rigid block can be characterized by the coincidence of the maximum rotational angle $\theta_{2\max}$ after the second impulse with the limit value $-\alpha$ of rotation. This condition gives the critical velocity amplitude of the double impulse just inducing the overturning of the rigid block. Note that the area A_2 in the restoring-force characteristic in the negative side can be obtained in closed form in terms of the velocity amplitude of the double impulse by using the conservation law of energy (Eq. (9.12)). In this process, the initial velocity of rotational angle was derived in closed form in Eq. (9.11) and the critical velocity amplitude limit of the double impulse can be obtained.

4. It was found from the explicit expression on the critical velocity amplitude limit of the double impulse that it is proportional to the square root of the size of the rigid block. This finding is important from the viewpoint of the structural design of monuments and tall buildings, i.e., as the structure becomes larger, it becomes more stable. This characteristic has been commonly understood and was confirmed here in a clear mathematical formulation.

5. Numerical examples including the comparison with the numerical simulation results by the Runge-Kutta method demonstrated the accuracy and reliability of the method explained in this chapter. However, the comparison of the response to the double impulse with that to the equivalent sinusoidal wave reminds us of the necessity to introduce a magnification coefficient (about 1.53–1.54 in this case). The introduction of two-step transformation of the magnitude of the equivalent sinusoidal waves (introduction of the one-cycle sinusoidal wave equivalent to the double impulse and magnification of the equivalent one-cycle sinusoidal wave) comes from the necessity of the equivalence of the maximum Fourier amplitude and from the viewpoint of connection to the previous work (Kojima and Takewaki 2016a).

6. The proposed magnified overturning limit exhibits a fairly good correspondence to the other available data for resonance (Makris and Kampas 2016, Dimitrakopoulos and DeJong 2012b).

Although only the critical input was dealt with in this chapter, it gives the lowest level of the limit input. If the input timing is not critical, it gives smaller responses (Taniguchi et al. 2017).

REFERENCES

Andreaus, U., and Casini, P. (1999). On the rocking-uplifting motion of a rigid block in free and forced motion: influence of sliding and impact, *Acta Mechanica*, 138, 219–241.

Anooshehpoor, A., Heaton, T. H., Shi, B., and Brune, J. N. (1999). Estimates of the ground accelerations at point Reyes station during the 1906 San Francisco earthquake, *Bull. Seismol. Soc. Am.*, 89, 845–853.

Casapulla, C., Jossa, P., and Maione, A. (2010). Rocking motion of a masonry rigid block under seismic actions: a new strategy based on the progressive correction of the resonance response, *Ingegneria Sismica*; 27(4): 35–48.

Casapulla, C., and Maione, A. (2016). Free damped vibrations of rocking rigid blocks as uniformly accelerated motions, *International J. of Structural Stability and Dynamics*, 10.1142/S0219455417500584.

Caughey, T. K. (1960a). Sinusoidal excitation of a system with bilinear hysteresis, *J. Appl. Mech.*, 27(4), 640–643.

Caughey, T. K. (1960b). Random excitation of a system with bilinear hysteresis, *J. Appl. Mech.*, 27(4), 649–652.

DeJong, M. J. (2012). Amplification of rocking due to horizontal ground motion, *Earthq. Spectra*, 28(4), 1405–1421.

Dimitrakopoulos, E. G., and DeJong, M. J. (2012a). Overturning of retrofitted rocking structures under pulse-type excitation, *J. Eng. Mech.*, ASCE, 138(8), 963–972.

Dimitrakopoulos, E. G., and DeJong, M. J. (2012b). Revisiting the rocking block: closed-form solutions and similarity laws, *Proc. R. Soc. A*, 468, 2294–2318.

ElGawady, M. A., Ma, Q., Butterworth, J. W., and Ingham, J. (2010). Effects of interface material on the performance of free rocking blocks, *Earthq. Eng. Struct. Dyn.*, 40, 375–392. (doi:10.1002/eqe.1025)

Hogan, S. J. (1989). On the dynamics of rigid-block motion under harmonic forcing, *Proc. R. Soc. Lond. A* 425, 441–476. (doi:10.1098/rspa.1989.0114)

Hogan, S. J. (1990). The many steady state responses of a rigid block under harmonic forcing, *Earthq. Eng. Struct. Dyn.*, 19, 1057–1071. (doi:10.1002/eqe.4290190709)

Housner, G. W. (1963). The behavior of inverted pendulum structures during earthquakes, *Bull. Seismol. Soc. Am.*, 53(2), 404–417.

Ishiyama, Y. (1982). Motions of rigid bodies and criteria for overturning by earthquake excitations, *Earthq. Eng. Struct. Dyn.*, 10, 635–650.

Iwan, W. D. (1961). *The dynamic response of bilinear hysteretic systems*, Ph.D. Thesis, California Institute of Technology, Pasadena.

Iwan, W. D. (1965a). The dynamic response of the one-degree-of-freedom bilinear hysteretic system, *Proc. of the Third World Conf. on Earthquake Eng.*, New Zealand.

Iwan, W. D. (1965b). The steady-state response of a two-degree-of-freedom bilinear hysteretic system, *J. Applied Mech.*, 32(1), 151–156.

Kojima, K., Fujita, K., and Takewaki, I. (2015). Critical double impulse input and bound of earthquake input energy to building structure, *Frontiers in Built Environment*, 1: 5.

Kojima, K., and Takewaki, I. (2015a). Critical earthquake response of elastic-plastic structures under near-fault ground motions (Part 1: Fling-step input), *Frontiers in Built Environment*, 1: 12.

Kojima, K., and Takewaki, I. (2015b). Critical earthquake response of elastic-plastic structures under near-fault ground motions (Part 2: Forward-directivity input), *Frontiers in Built Environment*, 1: 13.

Kojima, K., and Takewaki, I. (2015c). Critical input and response of elastic-plastic structures under long-duration earthquake ground motions, *Frontiers in Built Environment*, 1: 15.

Kojima, K., and Takewaki, I. (2016a). Closed-form critical earthquake response of elastic-plastic structures on compliant ground under near-fault ground motions, *Frontiers in Built Environment*, 2: 1.

Kojima, K., and Takewaki, I. (2016b). Closed-form dynamic stability criterion for elastic-plastic structures under near-fault ground motions, *Frontiers in Built Environment*, 2: 6.

Makris, N., and Black, C. J. (2004). Dimensional analysis of rigid-plastic and elasto-plastic structures under pulse-type excitations, *J. Eng. Mech.*, ASCE, 130, 1006–1018.

Makris, N., and Kampas, G. (2016). Size versus slenderness: Two competing parameters in the seismic stability of free-standing rocking columns, *Bull. Seismol. Soc. Am.*, 106(1) published online.

Milne, J. (1885). Seismic experiments, *Trans. Seism. Soc. Japan*, 8, 1–82.

Nabeshima, K., Taniguchi, R., Kojima, K., and Takewaki, I. (2016). Closed-form overturning limit of rigid block under critical near-fault ground motions, *Frontiers in Built Environment*, 2: 9.

Pompei, A., Scalia, A., and Sumbatyan, M. A. (1998). Dynamics of rigid block due to horizontal ground motion, *J. Eng. Mech. ASCE*, 124(7), 713–717.

Priestley, M. J. N., Evison, R. J., and Carr, A. J. (1978). Seismic response of structures free to rock on their foundations, *Bull. N.Z. Natl. Soc. Earthq. Eng.*, 11, 141–150.

Prieto, F., Lourenco, P. B., and Oliveira, C. S. (2004). Impulsive Dirac-delta forces in the rocking motion, *Earthq. Eng. Struct. Dyn.*, 33, 839–857. (doi:10.1002/eqe.381)

Shenton III, H. W., and Jones, N. P. (1991). Base excitation of rigid bodies: I Formulation, *J. Eng. Mech.*, ASCE, 117(10), 2286–2306.

Spanos, P. D. and Koh, A.-S. (1984). Rocking of rigid blocks due to harmonic shaking, *J. Eng. Mech.*, ASCE, 110(11), 1627–1642

Taniguchi, R., Nabeshima, K., Kojima, K., and Takewaki, I. (2017). Closed-form rocking vibration of rigid block under critical and non-critical double impulse, *Int. J. Earthquake and Impact Engineering*, 2(1), 32–45.

Yilmaz, C., Gharib, M., and Hurmuzlu, Y. (2009). Solving frictionless rocking block problem with multiple impacts, *Proc. R. Soc. A*, 465, 3323–3339. (doi:10.1098/rspa.2009.0273)

Yim, C. S., Chopra, A. K., and Penzien, J. (1980). Rocking response of rigid blocks to earthquakes, *Earthq. Eng. Struct. Dyn.*, 8(6), 565–587.

Zhang, J., and Makris, N. (2001). Rocking response of free-standing blocks under cycloidal pulses, *J. Eng. Mech.*, ASCE, 127(5), 473–483.

APPENDIX 1: VERIFICATION OF CRITICAL TIMING OF DOUBLE IMPULSE FOR VARIOUS INPUT LEVELS

In order to verify the critical timing of the double impulse, time-history response analysis was conducted for various input levels. By using the relation between the areas A_1 and A_2 in Figures 9.4 and 9.6 with the help of Eqs. (9.5), (9.6), (9.8), (9.10), (9.11), (9.12), and (9.15), the critical timing can be expressed by

$$t_{0c} = \frac{2}{p}\cosh^{-1}\left(\frac{1+\sqrt{r}}{\sqrt{\left(1+\sqrt{r}\right)^2 - \left(V/V_c\right)^2}}\right) \qquad (9.A1)$$

In Eq. (9.A1), V is an arbitrary input velocity, V_c is the critical overturning limit velocity obtained by Eq. (9.19), and a linear approximation $\sin(\alpha - \theta) \doteq (\alpha - \theta)$ was used.

Figure 9.A1 shows the plot of $\theta_{2\max}/\alpha$ with respect to t_0/t_{0c} where t_{0c} is given by Eq. (9.20) and t_0 is an arbitrary timing of the second impulse. It can be confirmed that the assumption introduced in Section 9.3 (critical timing is just after the impact) is valid.

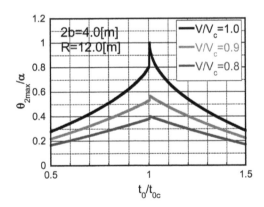

Figure 9.A1 Verification of critical timing: plot of $\theta_{2\max}/\alpha$ with respect to t_0/t_{0c} (Nabeshima et al. 2016).

APPENDIX J: VERIFICATION OF CRITICAL TIMING OF DOUBLE IMPULSE FOR VARIOUS INPUT LEVELS

Chapter 10

Critical earthquake response of a 2DOF elastic–perfectly plastic model under double impulse

10.1 INTRODUCTION

The double impulse is introduced here again as a substitute of the fling-step near-fault ground motion and a critical elastic-plastic response of a two-degree-of-freedom (2DOF) model under the "critical double impulse" is evaluated. The input of impulse at the base of building structures is expressed by the instantaneous change of velocities of the structural masses irrespective of the number of degrees of freedom of the building structures. Since only free-vibration appears under such double impulse, the energy balance approach plays an important role in the derivation of the solution of a complicated elastic-plastic critical response as in Kojima and Takewaki (2015a, b, c). However, phase lag occurs between the masses in the 2DOF model and a straightforward application of the energy balance approach for the SDOF model to the 2DOF model is difficult. It is shown that the critical timing of the double impulse is characterized by the timing of the second impulse at the zero story shear force in the first story. This criticality is also characterized by the maximization of the sum of the momenta of all masses. This timing certainly guarantees the maximum energy input by the second impulse which causes the maximum plastic deformation after the second impulse. As explained just above, because the response of 2DOF elastic-plastic building structures is quite complicated due to the phase lag between two masses compared to single-degree-of-freedom (SDOF) models for which a closed-form critical response can be derived, the upper bound of the critical response is introduced by using the convex model (Ben-Haim and Elishakoff 1990, Ben-Haim et al. 1996). The accuracy of the derived upper bound is discussed in comparison with the actual response analysis result to the double impulse. The validity and accuracy of the present theory for the double impulse are also investigated through the comparison with the result for the time-history response analysis to the corresponding one-cycle sinusoidal input as a representative of the fling-step near-fault ground motion.

10.2 DOUBLE IMPULSE INPUT

As in Chapters 2, 5, 7, 8, and 9, consider a double impulse ground acceleration $\ddot{u}_g(t)$, as shown in Figure 10.1, expressed by

$$\ddot{u}_g(t) = V\delta(t) - V\delta(t - t_0) \tag{10.1}$$

where V is the given velocity, $\delta(t)$ is the Dirac delta function and t_0 is the
time interval between two impulses. As in other chapters, the time derivative
is denoted by an over-dot.

The Fourier transform of $\ddot{u}_g(t)$ of the double impulse input can be
expressed as

$$\begin{aligned}
\ddot{U}_g(\omega) &= \int_{-\infty}^{\infty} \{V\delta(t) - V\delta(t - t_0)\} e^{-i\omega t} dt \\
&= \int_{-\infty}^{\infty} \{V\delta(t) e^{-i\omega t} - V\delta(t - t_0) e^{-i\omega t_0} e^{-i\omega(t - t_0)}\} dt \\
&= V(1 - e^{-i\omega t_0})
\end{aligned} \tag{10.2}$$

Figure 10.1 shows the acceleration and Fourier amplitude of the single
impulse and the double impulse. It should be remarked that the double
impulse possesses the periodicity. In addition, since the double impulse has

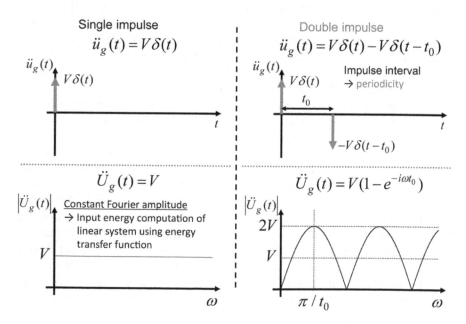

Figure 10.1 Single impulse and double impulse (acceleration and Fourier
amplitude).

power at higher frequencies different from the corresponding one-cycle sinusoidal wave, it should be careful in treating MDOF systems. Even if the fundamental natural frequency (or equivalent natural frequency of the elastic-plastic system) is resonant to the double impulse, higher natural modes may be induced by the double impulse.

10.3 TWO-DOF SYSTEM AND NORMALIZATION OF DOUBLE IMPULSE

Consider an undamped elastic–perfectly plastic (EPP) 2DOF model, as shown in Figure 10.2, subjected to the above-mentioned double impulse expressed by Eq. (10.1). Let m_i and k_i denote the mass and stiffness in the i-th story. The yield deformation and yield force of the i-th story are denoted by d_{yi} and f_{yi}. Furthermore, let ω_1, u_i, d_i, and f_i denote the undamped fundamental natural circular frequency of the EPP 2DOF system, the displacement of the i-th story mass relative to the ground, the interstory drift of the i-th story, and the restoring force of the i-th story, respectively. The velocity of the mass is denoted by $v_i = \dot{u}_i$.

Different from the situation in the case for the SDOF model, the reference value V_y of the velocity of the input double impulse is selected so that the input initial kinetic energy is transformed into the sum of the elastic limit strain energies.

$$\frac{1}{2}(m_1 + m_2)V_y^2 = \frac{1}{2}k_1d_{y1}^2 + \frac{1}{2}k_2d_{y2}^2 \qquad (10.3)$$

Although the interstory drifts of the first and second stories do not usually attain the elastic limit at the same time, this state is used only for normalizing the input level. Figure 10.3 shows an example of the time histories of the restoring forces in the first and second stories for a model of equal mass, equal story stiffness, and equal yield interstory drift subjected to the single

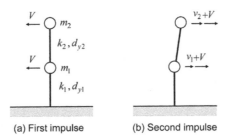

(a) First impulse   (b) Second impulse

Figure 10.2 Undamped EPP 2DOF model subjected to double impulse (Taniguchi et al. 2016).

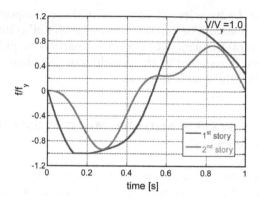

Figure 10.3 Example of time histories of restoring forces in first and second stories for model of equal mass, equal story stiffness, and equal yield interstory drift subjected to single impulse at $t=0$ with $V/V_y = 1.0$ (Taniguchi et al. 2016).

impulse at $t = 0$ with the input velocity amplitude $V/V_y=1$. The response was computed by using the Newmark-beta method.

10.4 DESCRIPTION OF ELASTIC-PLASTIC RESPONSE PROCESS IN TERMS OF ENERGY QUANTITIES

Let us describe the key phases in the response in terms of symbols as shown in Figure 10.4. Let $d^{(1)}_{i\max}$ and $d^{(2)}_{i\max}$ denote the maximum interstory drift after the first impulse and that after the second impulse, respectively, in the i-th

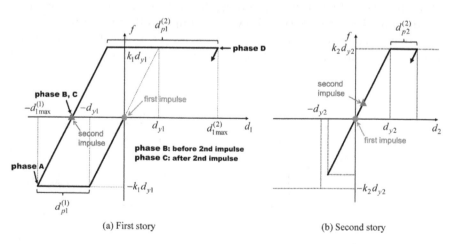

(a) First story

(b) Second story

Figure 10.4 Key phases (A), (B), (C), (D) in restoring-force characteristics: (a) first story, (b) second story (Taniguchi et al. 2016).

story. Furthermore, let $d_{pi}^{(1)}$ and $d_{pi}^{(2)}$ denote the plastic deformation after the first impulse and that after the second impulse, respectively, in the i-th story. The phase (A) indicates the state at which the first interstory drift attains the maximum value in the plastic region after the first impulse. On the other hand, the phase (B) presents the state just before the input of the second impulse and the phase (C) expresses the state just after the input of the second impulse. Furthermore, the phase (D) indicates the state at which the first interstory drift attains the maximum value in the counter-side plastic region after the second impulse. The phase (B) with zero first-story restoring force in Figure 10.4 was introduced because this timing plays an important role in Section 10.5.2.

In the phase (A), the velocity of the first story mass is zero. Let $E_2^{(A)}$ denote the sum of the elastic strain energy of the second story and the kinetic energy of the second-story mass. Although $E_2^{(A)}$ does not necessarily indicate the total mechanical energy of the second story, it is called the "total mechanical energy of the second story" for simplicity. In this case, the total mechanical energy $E_{sum}^{(A)}$ of the whole system at the phase (A) (the dissipated energy in the first story is not included because this is not the mechanical energy) can be expressed as

$$E_{sum}^{(A)} = \frac{1}{2} k_1 d_{y1}^2 + E_2^{(A)} \tag{10.4}$$

Since the mechanical energy is conserved in the process from the phase (A) to the phase (B) ($E_{sum}^{(B)}$: total mechanical energy at the phase (B)), the following relation holds.

$$E_{sum}^{(B)} = E_{sum}^{(A)} \tag{10.5}$$

Let ΔE denote the energy input caused by the second impulse. In this case, the total mechanical energy $E_{sum}^{(C)}$ at the phase (C) can be related to $E_{sum}^{(B)}$ as

$$E_{sum}^{(C)} = E_{sum}^{(B)} + \Delta E \tag{10.6}$$

Let $E_2^{(D)}$ denote the sum of the elastic strain energy of the second story and the kinetic energy of the second-story mass at the phase (D). In this case, the total mechanical energy $E_{sum}^{(D)}$ at the phase (D) can be related to $E_2^{(D)}$ as

$$E_{sum}^{(D)} = \frac{1}{2} k_1 d_{y1}^2 + E_2^{(D)} \tag{10.7}$$

The energy balance provides

$$E_{sum}^{(C)} = E_{sum}^{(D)} + k_1 d_{y1} d_{p1}^{(2)} + k_2 d_{y2} d_{p2}^{(2)} \tag{10.8}$$

It may be rare that the second story goes into the plastic region after the second impulse. This issue will be discussed later in Section 10.6.3. Therefore, $d_{p2}^{(2)}=0$ holds in most cases. Substitution of Eqs. (10.4), (10.5), (10.7), and (10.8) into Eq.(10.6) provides

$$\frac{1}{2}k_1 d_{y1}^2 + E_2^{(A)} + \Delta E = \frac{1}{2}k_1 d_{y1}^2 + E_2^{(D)} + k_1 d_{y1} d_{p1}^{(2)} + k_2 d_{y2} d_{p2}^{(2)} \quad (10.9)$$

From Eq. (10.9) the normalized plastic deformation of the first story after the second impulse can be expressed by

$$\frac{d_{p1}^{(2)}}{d_{y1}} = \frac{1}{k_1 d_{y1}^2}\left\{E_2^{(A)} + \Delta E - \left(E_2^{(D)} + k_2 d_{y2} d_{p2}^{(2)}\right)\right\} \quad (10.10)$$

Eq. (10.10) indicates that the upper bound of the plastic deformation $d_{p1}^{(2)}$ of the first story after the second impulse is derived by maximizing $E_2^{(A)}$ and ΔE and minimizing $E_2^{(D)}$ and $d_{p2}^{(2)}$. These manipulations will be discussed in the following. Since the undamped EPP 2DOF model includes some uncertain factors in the state determination different from SDOF models, the investigation on upper bound of responses for variable uncertain parameters may be meaningful (Takewaki 1996, 1997, 2002).

Only critical response is taken into account by the present method and the critical resonant frequency can be obtained without iteration for the increasing input level as shown in Section 10.5. One of the original points in this chapter is the introduction of the concept of "critical excitation" in the elastic-plastic response for multi-degree-of-freedom (MDOF) models (Drenick 1970, Abbas and Manohar 2002, Takewaki 2007, Moustafa et al. 2010). Once the frequency and amplitude of the critical double impulse are computed, the corresponding one-cycle sinusoidal motion can be identified (see Section 1.2 in Chapter 1).

10.5 UPPER BOUND OF PLASTIC DEFORMATION IN FIRST STORY AFTER SECOND IMPULSE

10.5.1 Maximization of $E_2^{(A)}$

Assume the situation after the first impulse where the first story is in the plastic loading range and the second story is in the elastic range. This case is often encountered in usual situations (for example, Bertero et al. 1978) and some examples will be shown later for the special model of equal mass, equal story stiffness, and equal yield interstory drift. In this case, the equations of motion after the yielding of the first story, i.e., until phase (A), can be expressed by

$$m_1\ddot{u} - k_2(u_2 - u_1) = k_1 d_{y1} \quad\quad\quad (10.11\text{a})$$

$$m_2\ddot{u}_2 + k_2(u_2 - u_1) = 0 \quad\quad\quad (10.11\text{b})$$

Arrangement of Eqs. (10.11a, b) yields

$$m_1 m_2(\ddot{u}_2 - \ddot{u}_1) + k_2(m_1 + m_2)(u_2 - u_1) = -m_2 k_1 d_{y1} \quad\quad\quad (10.12)$$

The general solution of the differential equation, Eq. (10.12), can be expressed by

$$u_2 - u_1 = -B\cos(\omega t + \delta) - A \quad\quad\quad (10.13)$$

where B is an undetermined coefficient and

$$\omega = \sqrt{\frac{(m_1 + m_2)k_2}{m_1 m_2}}, \quad A = \frac{m_2 k_1}{(m_1 + m_2)k_2} d_{y1} \quad\quad\quad (10.14\text{a,b})$$

Eq. (10.13) indicates that the second-story interstory drift $u_2 - u_1$ exhibits a simple harmonic vibration around the center of magnitude A as shown in Figure 10.5. Since the first story goes into the plastic range quickly and the second-story interstory drift has a zero initial value, the absolute minimum value of the second-story interstory drift is nearly zero. It can be shown that this assumption is a good approximation.

In this case, the following relation holds.

$$B \doteq A \quad\quad\quad (10.15)$$

$$u_2 - u_1 \doteq -A\{\cos(\omega t + \delta) + 1\} \quad (\delta: \text{ phase angle}) \quad\quad\quad (10.16)$$

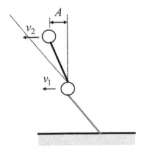

Figure 10.5 Simple harmonic vibration of second-story mass around center of magnitude A (Taniguchi et al. 2016).

Then, the condition that the second story does not go into the plastic range after the first impulse can be expressed by

$$2A = \frac{2m_2 k_1}{(m_1 + m_2)k_2} d_{y1} \leq d_{y2} \qquad (10.17)$$

The model of equal mass, equal story stiffness, and equal yield deformation satisfies this condition.

Some examples are shown in Figure 10.6 for the time histories of the restoring forces in the first and second stories for the model of equal mass, equal story stiffness, equal yield interstory drift subjected to the single impulse at $t = 0$ with $V/V_y=1.15$, 1.5, 2, 4. As in the previous numerical example, the response was computed by using the Newmark-beta method. These numerical investigations (Figure 10.6) enable the derivation of the following approximate relation.

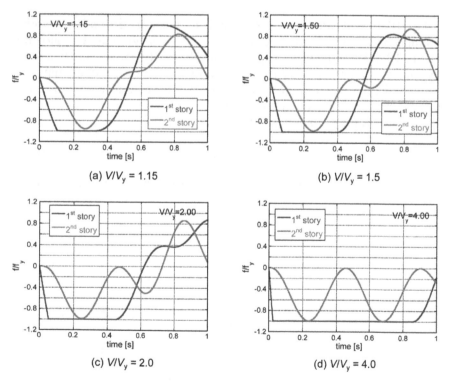

Figure 10.6 Examples of time histories of restoring forces in first and second stories for model of equal mass, equal story stiffness, and equal yield interstory drift subjected to single impulse at $t = 0$ with $V/V_y=1.15$, 1.5, 2, 4 (Taniguchi et al. 2016).

$$\frac{m_2 k_1}{(m_1 + m_2) k_2} d_{y1} = \frac{1}{2} d_{y1} \tag{10.18}$$

Eq. (10.16) can also be expressed by

$$\dot{u}_2 - \dot{u}_1 \doteq A\omega \sin(\omega t + \delta) \tag{10.19}$$

Since the first story mass is at rest in the phase (A), the velocity of the second-story mass can be described by

$$\dot{u}_2 = A\omega \sin(\omega t + \delta) \tag{10.20}$$

For simplicity of expression, let us introduce the notation $\mu = m_2/m_1$, $\kappa = k_2/k_1$. The total mechanical energy in the second story at the phase (A) can then be derived by

$$E_2^{(A)} = \frac{1}{2} k_2 (u_2 - u_1)^2 + \frac{1}{2} m_2 \dot{u}_2^2 \tag{10.21}$$

Substitution of Eqs. (10.16) and (10.20) into Eq. (10.21) leads to

$$
\begin{aligned}
E_2^{(A)} &= \frac{1}{2} A^2 \left[k_2 \{\cos(\omega t + \delta) + 1\}^2 + m_2 \omega^2 \sin^2(\omega t + \delta) \right] \\
&= \frac{1}{2} k_2 A^2 \left[\{\cos(\omega t + \delta) + 1\}^2 + (\mu + 1)\sin^2(\omega t + \delta) \right] \tag{10.22} \\
&= \frac{1}{2} k_2 \left\{ \frac{\mu}{\mu + 1} \frac{1}{\kappa} d_{y1} \right\}^2 \left[-\mu \left\{ \cos(\omega t + \delta) - \frac{1}{\mu} \right\}^2 + \frac{1}{\mu}(\mu + 1)^2 \right]
\end{aligned}
$$

In Eq. (10.22), the minimization of $\{\cos(\omega t + \delta) - (1/\mu)\}^2$ leads to the maximization of $E_2^{(A)}$. Since $0 \leq |\cos(\omega t + \delta)| \leq 1$, it is necessary to consider two cases $\mu > 1$ and $\mu \leq 1$. In the case of $\mu > 1$, the condition $\cos(\omega t + \delta) - (1/\mu) = 0$ minimizes $\{\cos(\omega t + \delta) - (1/\mu)\}^2$. On the other hand, in case of $\mu \leq 1$, the condition $\cos(\omega t + \delta) = 1$ minimizes $\{\cos(\omega t + \delta) - (1/\mu)\}^2$.

Finally, the following relations can be drawn.

$$0 \leq E_2^{(A)} \leq E_{2\max}^{(A)} \tag{10.23}$$

where $E_{2\max}^{(A)}$ is the upper bound of $E_2^{(A)}$ and is given by

$$E_{2\max}^{(A)} = \frac{\mu}{2\kappa} k_1 d_{y1}^2 \quad (\mu > 1) \tag{10.24a}$$

$$E_{2\max}^{(A)} = \frac{2\mu^2}{(\mu+1)^2 \kappa} k_1 d_{y1}^2 \quad (\mu \le 1) \tag{10.24b}$$

10.5.2 Maximization of ΔE (minimization of ΔE in addition)

Since the displacements of masses do not change at once at the action of the second impulse, the increment (change) of the total mechanical energy just before and after the second impulse, i.e., the total energy input, can be expressed by the change of kinetic energies (see Figure10.7(a)).

$$\Delta E = \frac{1}{2} m_1 (v_1 + V)^2 + \frac{1}{2} m_2 (v_2 + V)^2 - \frac{1}{2} m_1 v_1^2 - \frac{1}{2} m_2 v_2^2$$
$$= (m_1 v_1 + m_2 v_2) V + \frac{1}{2} (m_1 + m_2) V^2 \tag{10.25}$$

Eq. (10.25) indicates that the timing of the second impulse at the maximum sum of the momenta $P = m_1 v_1 + m_2 v_2$ actually maximizes the total energy input ΔE. Since the time derivative of the sum of the momenta is equal to the force $f_1(t)$ in the first story, the following relation holds.

$$\frac{dP}{dt} = f_1(t) \tag{10.26}$$

Therefore, the maximization of the sum of the momenta can be characterized by (see Figure 10.7(b))

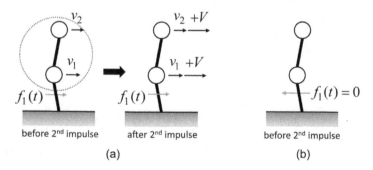

before 2nd impulse   after 2nd impulse   before 2nd impulse

(a)   (b)

Figure 10.7 Undamped EPP 2DOF model under double impulse, (a) story shear force $f_1(t)$ in first story and velocities of masses, (b) critical timing of second impulse.

$$f_1(t) = 0 \qquad (10.27)$$

Substitution of the condition $d_1 - d_{r1} = 0$ (d_{r1}: residual deformation in the first story), derived from Eq. (10.27), into the expression of the total mechanical energy at the phase (B) leads to the equation of the energy conservation between the phase (B) and (A).

$$\frac{1}{2}m_1v_1^2 + \frac{1}{2}m_2v_2^2 + \frac{1}{2}k_2d_2^2 = \frac{1}{2}k_1d_{y1}^2 + E_2^{(A)} \qquad (10.28)$$

Rearrangement of Eq. (10.28) provides

$$\frac{1}{2}m_1v_1^2 + \frac{1}{2}m_2v_2^2 = \frac{1}{2}k_1d_{y1}^2 + E_2^{(A)} - \frac{1}{2}k_2d_2^2 \qquad (10.29)$$

Since the left-hand side of Eq. (10.29) is positive, the following inequality can be derived.

$$0 \le \frac{1}{2}k_2d_2^2 \le \frac{1}{2}k_1d_{y1}^2 + E_2^{(A)} \le \frac{1}{2}k_1d_{y1}^2 + E_{2\,\text{max}}^{(A)} \qquad (10.30)$$

Let introduce the following quantity D.

$$D = \frac{1}{2}k_1d_{y1}^2 + E_{2\,\text{max}}^{(A)} \qquad (10.31)$$

Eq. (10.29) and the positivity of $k_2d_2^2/2$ lead to the following relation (see Figure 10.8).

$$\frac{1}{2}m_1v_1^2 + \frac{1}{2}m_2v_2^2 \le D \qquad (10.32)$$

The tangential line of the ellipse $\left(m_1v_1^2/2\right) + \left(m_2v_2^2/2\right) = D$ at the point (v_{10}, v_{20}) can be expressed by

$$\frac{v_{10}}{m_2}v_1 + \frac{v_{20}}{m_1}v_2 = \frac{2D}{m_1m_2} \qquad (10.33)$$

Recalling $P = m_1v_1 + m_2v_2$, Eq. (10.33) provides

$$\frac{v_{10}}{m_2} : \frac{v_{20}}{m_1} : \frac{2D}{m_1m_2} = m_1 : m_2 : P, \qquad (10.34)$$

The first proportionality relation in Eq. (10.34) yields $v_{10} = v_{20}$. The substitution of this relation into Eq. (10.33) and the other proportionality relation in Eq. (10.34) leads to

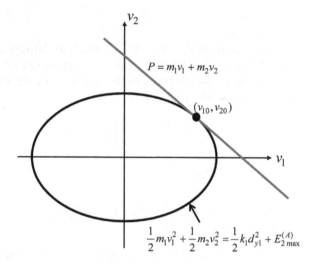

Figure 10.8 Tangential line of ellipse $\left(m_1v_1^2/2\right)+\left(m_2v_2^2/2\right)=D$ at point $(v_{10},\ v_{20})$ (Taniguchi et al. 2016).

$$P = \sqrt{2\left(m_1 + m_2\right)D} \tag{10.35}$$

Finally, the upper bound $\Delta E_{\max}$ of the input energy at the second impulse can be derived as

$$\Delta E_{\max} = \frac{1}{2}\left(m_1 + m_2\right)V^2 + V\sqrt{2\left(m_1 + m_2\right)\left(\frac{1}{2}k_1d_{y1}^2 + E_{2\max}^{(A)}\right)} \tag{10.36}$$

This process employs the application of the convex model (Ben-Haim and Elishakoff 1990). It may be interesting to note that the property (Eqs. (10.26) and (10.27)) and the proof shown in this section can be extended to multi-degree-of-freedom (MDOF) models.

On the other hand, it is meaningful to derive the lower bound of the input energy by the second impulse even approximately because the upper bound may not be a tight bound.

Assume that the total mechanical energy at the phase (A) is transformed into the kinetic energy of the first story mass and that the second story does not have the strain energy and the kinetic energy. This assumption can be expressed by

$$\frac{1}{2}m_1v_1^2 = \frac{1}{2}k_1d_{y1}^2 + E_2^{(A)} \tag{10.37}$$

Eq. (10.37) leads to

$$v_1 = \sqrt{\frac{k_1}{m_1} d_{y1}^2 + \frac{2E_2^{(A)}}{m_1}}$$ (10.38)

$$P = \sqrt{m_1 \left(k_1 d_{y1}^2 + 2E_2^{(A)}\right)}$$ (10.39)

Therefore, the approximate lower bound $\Delta E_{\min}$ of the input energy by the second impulse can be derived as

$$\Delta E_{\min} = \frac{1}{2}(m_1 + m_2)V^2 + V\sqrt{m_1\left(k_1 d_{y1}^2 + 2E_2^{(A)}\right)}$$ (10.40)

10.5.3 Minimization of $E_2^{(D)} + k_2 d_{y2} d_{p2}^{(2)}$ (maximization of $E_2^{(D)} + k_2 d_{y2} d_{p2}^{(2)}$ in addition)

Since $E_2^{(D)}$ is the sum of the second-story strain energy and the second-story kinetic energy, the following relation can be derived by minimizing $E_2^{(D)} + k_2 d_{y2} d_{p2}^{(2)}$.

$$E_{2\min}^{(D)} = 0$$ (10.41)

$$d_{p2\min}^{(2)} \doteq 0$$ (10.42)

On the other hand, the maximization of $E_2^{(D)}$ may be difficult. Because the effect of the second impulse on the second-story interstory drift is small, it may be reasonable to assume that $d_{p2\max}^{(2)} \doteq 0$. Therefore, the following relations can be born approximately.

$$E_{2\max}^{(D)} \doteq \frac{1}{2}k_2 d_{y2}^2$$ (10.43)

$$d_{p2\max}^{(2)} \doteq 0$$ (10.44)

10.5.4 Upper bound of plastic deformation in the first story after the second impulse

Based on the results of Sections 10.5.1, 10.5.2, and 10.5.3, the upper bound of $d_{p1}^{(2)}$ can be derived by

$$\frac{d_{p1}^{(2)}}{d_{y1}} \le \frac{1}{k_1 d_{y1}^2}\left\{\frac{\mu}{2\kappa}k_1 d_{y1}^2 + \frac{1}{2}(m_1+m_2)V^2\right.$$

$$+ \sqrt{(m_1+m_2)\left(1+\frac{\mu}{\kappa}\right)k_1 d_{y1}^2 V}\right\}$$

$$= \frac{1}{k_1 d_{y1}^2}\left\{\frac{\mu}{2\kappa}k_1 d_{y1}^2 + a^2\left(\frac{1}{2}k_1 d_{y1}^2 + \frac{1}{2}k_2 d_{y2}^2\right)\right.$$

$$\left. +a\sqrt{\left(1+\frac{\mu}{\kappa}\right)(k_1 d_{y1}^2 + k_2 d_{y2}^2)k_1 d_{y1}^2}\right\} \quad (\mu > 1) \tag{10.45a}$$

$$\frac{d_{p1}^{(2)}}{d_{y1}} \le \frac{1}{k_1 d_{y1}^2}\left\{\frac{2\mu^2}{(\mu+1)^2\kappa}k_1 d_{y1}^2 + \frac{1}{2}(m_1+m_2)V^2\right.$$

$$+ \sqrt{(m_1+m_2)\left(1+\frac{4\mu^2}{(\mu+1)^2\kappa}\right)k_1 d_{y1}^2 V}\right\}$$

$$= \frac{1}{k_1 d_{y1}^2}\left\{\frac{2\mu^2}{(\mu+1)^2\kappa}k_1 d_{y1}^2 + a^2\left(\frac{1}{2}k_1 d_{y1}^2 + \frac{1}{2}k_2 d_{y2}^2\right)\right.$$

$$\left. +a\sqrt{\left(1+\frac{4\mu^2}{(\mu+1)^2\kappa}\right)(k_1 d_{y1}^2 + k_2 d_{y2}^2)k_1 d_{y1}^2}\right\} \quad (\mu \le 1) \tag{10.45b}$$

In the case of the model of equal mass ($\mu = 1$), equal story stiffness and equal yield interstory drift, Eq.(10.45b) can be reduced to

$$\frac{d_{p1}^{(2)}}{d_{y1}} \le a^2 + 2a + \frac{1}{2} \quad (a = V/V_y) \tag{10.46}$$

On the other hand, an approximate lower bound of $d_{p1}^{(2)}$ can be derived by substituting Eqs. (10.23), (10.40), (10.43), and (10.44) into Eq.(10.10).

$$\frac{d_{p1}^{(2)}}{d_{y1}} \ge \frac{1}{k_1 d_{y1}^2}\left\{\frac{1}{2}(m_1+m_2)V^2 + \sqrt{m_1 k_1 d_{y1}^2}V - \frac{1}{2}k_2 d_{y2}^2\right\}$$

$$= \frac{1}{k_1 d_{y1}^2}\left\{\frac{a^2}{2}(k_1 d_{y1}^2 + k_2 d_{y2}^2) + a\sqrt{\frac{m_1}{m_1+m_2}(k_1 d_{y1}^2 + k_2 d_{y2}^2)k_1 d_{y1}^2} - \frac{1}{2}k_2 d_{y2}^2\right\} \tag{10.47}$$

In the case of the model of equal mass, equal story stiffness, and equal yield interstory drift, Eq. (10.47) can be reduced to

$$\frac{d_{p1}^{(2)}}{d_{y1}} \ge a^2 + a - \frac{1}{2} \quad (a = V/V_y) \tag{10.48}$$

10.6 NUMERICAL EXAMPLES OF CRITICAL RESPONSES

10.6.1 Upper bound of critical response

Consider an undamped EPP 2DOF model of equal mass ($m_1 = m_2$=1.0×10⁶kg), equal story stiffness ($k_1 = k_2$=1.0×10⁸N/m), and equal yield interstory drift ($d_{y1} = d_{y2}$=0.1m) subjected to the double impulse. Since this 2DOF model is a simplified model of a multi-story model (about 10-story), a rather large yield interstory drift (sum of each story's yield interstory drift) was assumed. The fundamental natural period is 1.02(s). For comparison, a reduced SDOF model as shown in Figure10.9 is considered.

Figure10.10 shows the maximum interstory drift in the first story after the first impulse. In this figure, the response of the SDOF system as shown in Figure10.9 (also upper bound of 2DOF system) and the response under the corresponding one-cycle sinusoidal wave are also plotted. The amplitudes of the double impulse and the corresponding one-cycle sinusoidal wave are adjusted so as to have the same maximum Fourier amplitude (Section 1.2 in Chapter 1 and Appendix 1). It should be remarked that the assumption of energy concentration into the first story provides the upper bound of the EPP 2DOF system and this assumption is equivalent to the modeling into the reduced SDOF system. The response was computed by using the Newmark-beta method. It can be observed that the upper bound of the EPP 2DOF system can bound the actual critical response tightly. Although the response under the corresponding one-cycle sinusoidal wave is rather small in a larger input level, the correspondence up to about $V/V_y = 3$ may be meaningful from the practical viewpoint.

Figure 10.11 presents the maximum interstory drift after the second impulse in the first story in which the response of the SDOF system and the response under the corresponding one-cycle sinusoidal wave are also plotted. The second impulse was acted at the zero-restoring-force timing in the first story (Eq. (10.27)) in the Newmark-beta method. As stated above, the amplitudes of the double impulse and the corresponding one-cycle

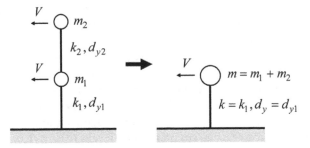

Figure 10.9 Transformation of 2DOF system into reduced SDOF system (Taniguchi et al. 2016).

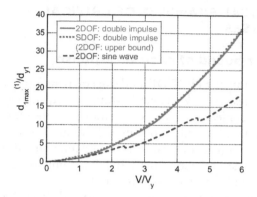

Figure 10.10 Maximum interstory drift after first impulse in first story together with response of SDOF system (also upper bound of 2DOF system) and response under corresponding one-cycle sinusoidal wave (Taniguchi et al. 2016).

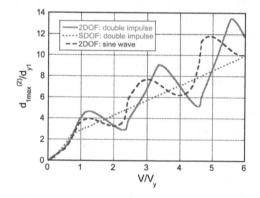

Figure 10.11 Maximum interstory drift after second impulse in first story (Taniguchi et al. 2016).

sinusoidal wave are adjusted so as to have the same maximum Fourier amplitude (Section 1.2 in Chapter 1). In Figure 10.11, the correspondence up to about $V/V_y = 3$ may be meaningful from the practical viewpoint (the level of ductility factor).

Figure 10.12 shows the critical plastic deformation after the second impulse in the first story and their upper and lower bounds (Eqs. (10.46) and (10.48)) in which the upper bound of the SDOF model and the response under the corresponding one-cycle sinusoidal wave are also plotted. It should be noted that in the case of the elastic deformation in the first story after the first impulse (rather smaller input level: $V/V_y \leq 1.0$ in this case), another expressions of the upper and lower bounds shown in Appendix 2 have to be employed.

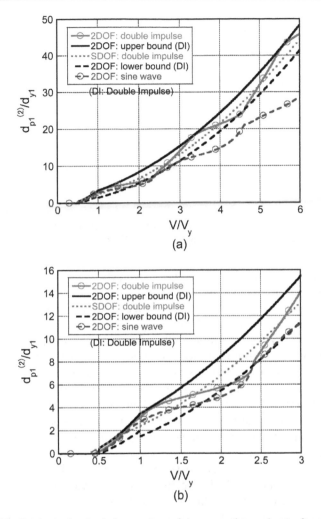

Figure 10.12 Critical plastic deformation after second impulse in first story and their upper and lower bounds together with upper bound of SDOF system and response under corresponding one-cycle sinusoidal wave: (a) comparison up to $V/V_y = 6$, (b) magnified (Taniguchi et al. 2016).

It can be understood that the actual critical plastic deformation after the second impulse in the first story corresponds fairly well with the response under the corresponding one-cycle sinusoidal wave up to about $V/V_y = 3$ of the input velocity level. Furthermore, the upper and lower bounds can bound the actual critical plastic deformation (although the lower bound is approximate due to the uncertain assumption of Eq. (10.44)).

Figure 10.13 presents the critical timing t_{0c} of the second impulse with respect to the input level of the double impulse. This critical timing was

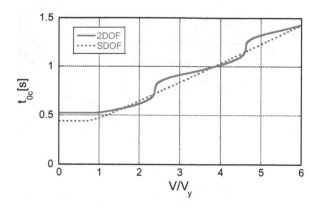

Figure 10.13 Critical timing of second impulse with respect to input level of double impulse (Taniguchi et al. 2016).

obtained using Eq. (10.27). For comparison, the critical timing for the corresponding SDOF model is also plotted. Consider that a structure is given, i.e., the parameters of the structure are specified. Then the critical timing t_{0c} can be found from Figure10.13 (although the critical timing t_{0c} of a 2DOF model has to be evaluated numerically). The practical range of the period of pulses in near-fault ground motions may be 0.5s-3s. The fundamental natural period of most of buildings is in this range except very flexible high-rise buildings and base-isolated buildings. An important matter in the seismic design of structures for near-fault ground motions is to take into account the most unfavorable situation (resonant in elastic and elastic-plastic range). This can be justified because earthquake ground motions are highly uncertain (Takewaki 2007).

It may be interesting to demonstrate the correspondence of the response to the double impulse with that to the sinusoidal wave in the time domain. The comparison of the time histories (first-story interstory drift and first-story restoring force) and the first-story restoring-fore characteristic under the double impulse and the corresponding sinusoidal wave is shown in Appendix 3.

10.6.2 Input level for tight upper bound

Consider the same undamped EPP 2DOF model subjected to the double impulse with V/V_y=1.11, 3.33, 5.55 at which the upper bound is close to the actual critical response in Figure 10.12. The increment 2.22 of the input level was analyzed by using the fundamental law in dynamics (Newton's second law) (see Appendix 4). Figure 10.14 presents the time histories of the restoring forces in the first and second stories for the same 2DOF model subjected to the double impulse with V/V_y=1.11, 3.33, 5.55. The critical timing t_{0c} is also shown in the figure captions.

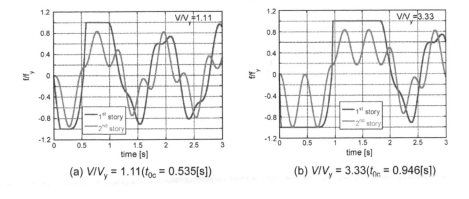

(a) $V/V_y = 1.11 (t_{0c} = 0.535[s])$ (b) $V/V_y = 3.33 (t_{0c} = 0.946[s])$

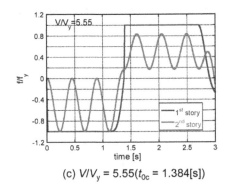

(c) $V/V_y = 5.55 (t_{0c} = 1.384[s])$

Figure 10.14 Time histories of restoring forces in first and second stories for the same 2DOF model subjected to double impulse with $V/V_y=1.11$, 3.33, 5.55: (a) $V/V_y=1.11$ ($t_{0c}=0.535$ [s]), (b) $V/V_y= 3.33$ ($t_{0c}=0.946$ [s]), (c) $V/V_y=5.55$ ($t_{0c}=1.384$[s]) (Taniguchi et al. 2016).

Figure 10.15 shows the deformation mode and velocity at the phase (A). In addition, Figure 10.16 illustrates the time histories of the restoring forces in the first and second stories starting from the phase (A) (only the first impulse is given: $V/V_y=1.11$, 3.33, 5.55). Furthermore, Figure 10.17 presents the velocity (first-story mass)-velocity (second-story mass) plane for a variable motion of masses attaining the maximum sum of momenta (tangential line) by the convex model and the actual motion of masses starting from the phase (A) (only the first impulse is given: $V/V_y=1.11$, 3.33, 5.55). Figure 10.17 indicates that the actual motion of masses passes through the response derived by the convex model.

10.6.3 Input level for loose upper bound

Consider the same undamped EPP 2DOF model subjected to the double impulse with the input velocity levels $V/V_y=2.22$, 4.44 at which the upper

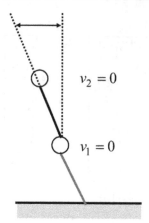

Figure 10.15 Deformation mode and velocity at phase (A) (upper bound case) (Taniguchi et al. 2016).

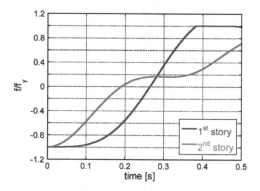

Figure 10.16 Time histories of restoring forces in first and second stories starting from phase (A) (Taniguchi et al. 2016).

bound is far from the actual critical response in Figure 10.12 (the lower bound is close to the actual critical one). As stated in Section 10.6.2, the increment 2.22 of the input level was analyzed by using the fundamental law in dynamics (Newton's second law). Figure 10.18 shows the time histories of the restoring forces in the first and second stories for the same 2DOF model subjected to the double impulse with V/V_y=2.22, 4.44. In this example, $d_{p2}^{(2)}$ exists which was discussed in Eq. (10.8). The critical timing t_{0c} is also shown in the figure (t_{0c}=0.658 [s] for V/V_y= 2.22 and t_{0c}=1.089 [s] for V/V_y= 4.44).

Figure 10.19 shows the deformation mode and velocity at the phase (A). Figure 10.20 presents the time histories of the restoring forces in the first and second stories starting from the phase (A) (only the first impulse is given: V/V_y=2.22, 4.44). Figure 10.21 illustrates the velocity (first-story

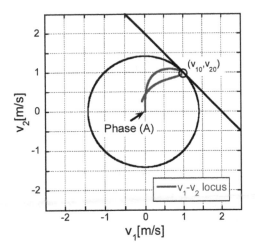

Figure 10.17 Velocity (first-story mass)-velocity (second-story mass) plane for variable motion of masses attaining maximum sum of momenta (tangential line) by convex model and actual motion of masses starting from phase (A) (only first impulse is given) (Taniguchi et al. 2016).

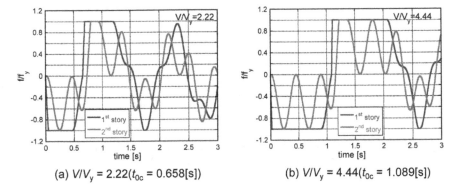

(a) $V/V_y = 2.22(t_{0c} = 0.658[s])$

(b) $V/V_y = 4.44(t_{0c} = 1.089[s])$

Figure 10.18 Time histories of restoring forces in first and second stories for the same 2DOF model subjected to double impulse with V/V_y=2.22, 4.44: (a) V/V_y= 2.22 (t_{0c}=0.658 [s]), (b) V/V_y= 4.44 (t_{0c}=1.089 [s]) (Taniguchi et al. 2016).

mass)-velocity (second-story mass) plane for a variable motion of masses bounded by the shrunk circle and the actual motion of masses starting from the phase (A) (only the first impulse is given: V/V_y=2.22, 4.44). Figure 10.21 clearly shows that the actual motion of masses is far from the solution derived by the convex model.

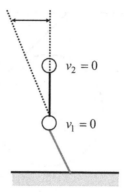

Figure 10.19 Deformation mode and velocity at phase (A) (lower bound case) (Taniguchi et al. 2016).

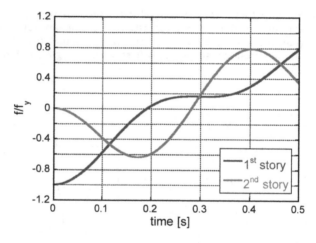

Figure 10.20 Time histories of restoring forces in first and second stories starting from phase (A) (Taniguchi et al. 2016).

10.6.4 Verification of criticality

The criticality of the timing of the second impulse defined by Eq. (10.27) is investigated in this section. For this purpose, a parametric analysis for the varied timing of the second impulse is performed. Figure 10.22 shows the plot of $d_{p1}^{(2)}/d_{y1}$ with respect to the timing of the second impulse. The restoring force in the first story at the second impulse is also plotted for reference. We can see that the criterion of the zero restoring force in the first story defined by Eq. (10.27) certainly maximizes $d_{p1}^{(2)}/d_{y1}$ in addition to maximizing the input energy at the second impulse.

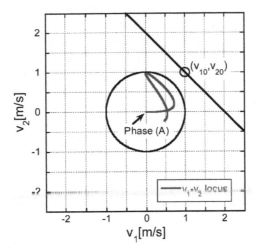

Figure 10.21 Velocity (first-story mass)-velocity (second-story mass) plane for variable motion of masses bounded by shrunk circle and actual motion of masses starting from phase (A) (only first impulse is given) (Taniguchi et al. 2016).

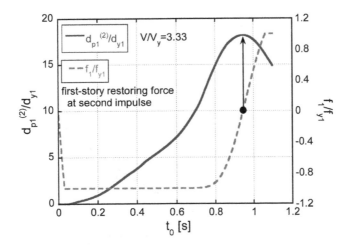

Figure 10.22 Plot of $d_{p1}^{(2)}/d_{y1}$ and f_1/f_{y1} (restoring force in first story at second impulse) with respect to timing of second impulse (Taniguchi et al. 2016).

10.7 APPLICATION TO RECORDED GROUND MOTIONS

To investigate the applicability of the present theory to actual recorded pulse-type ground motions, a comparison of the proposed upper bound with the corresponding response to a recorded ground motion is presented.

Consider the Rinaldi station fault-normal component during the Northridge earthquake in 1994 as a representative pulse-type ground motion. The structural model is the same as in Section 10.6.1 (equal mass, equal stiffness, and equal yield interstory drift in each story). Since the ground motion is fixed (V and t_0 are fixed), the structural model parameters are varied, i.e., V_y (specifically $k_1 = k_2$ and $d_{y1} = d_{y2}$) is varied. Figure 10.23(a) illustrates the modeling of a part of the recorded ground motion acceleration into a one-cycle sinusoidal input. Figure 10.23(b) shows the maximum amplitude of plastic deformation for the recorded ground motion by time-history response analysis and the proposed upper bound for the corresponding double impulse. As stated before, since the initial velocity V is determined in Figure 10.23(a), V_y is changed here. This procedure is similar to the well-known elastic-plastic response spectrum developed in 1960–1970. The solid

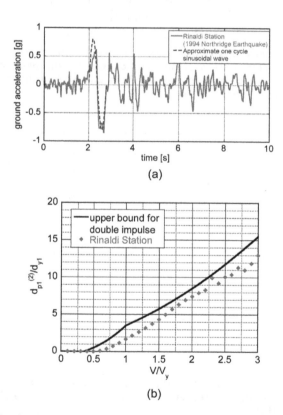

Figure 10.23 Applicability to recorded ground motion: (a) modeling of part of pulse-type recorded ground motion (Rinaldi station fault-normal component during the Northridge earthquake in 1994) into corresponding one-cycle sinusoidal input; (b) maximum amplitude of plastic deformation for recorded ground motion and proposed upper bound for corresponding double impulse (Taniguchi et al. 2016).

line is obtained by changing V_y for the specified V using the proposed method for the upper bound to the double impulse, and the dotted line is drawn by conducting the elastic-plastic time-history response analysis on each model with varied V_y under the recorded ground motion. It can be observed that the result by the proposed method is a fairly good upper-bound approximate of the result to the recorded pulse-type ground motion.

10.8 SUMMARIES

The conclusions may be summarized as follows:

1. The critical elastic-plastic response of an undamped EPP 2DOF model under the "critical double impulse" was evaluated. The critical excitation problem is such that the velocity amplitude of the double impulse is fixed and the interval of the double impulse is variable. Since only free-vibration appears under such double impulse, the energy balance approach plays a critical role in the direct derivation of the complicated elastic-plastic critical response.
2. The criticality is characterized by the timing of the second impulse at the zero story shear force in the first story. This timing guarantees the maximum energy input by the second impulse, which is expected to cause the maximum plastic deformation after the second impulse in the first story. This critical timing also coincides with the state in which the sum of the momenta attains the maximum. This property can be extended to multi-degree-of-freedom models (see Chapter 11).
3. Because the response of an undamped EPP 2DOF model is quite complicated due to the phase lag between two masses compared to the SDOF model for which a closed-form critical response can be derived, the upper bound of the critical response was investigated. The upper bound was derived by using the convex model partially, which is a powerful tool for uncertainty analysis.
4. As for the maximum interstory drift in the first story after the first impulse, the derived upper bound is close to the actual maximum response evaluated by time-history response analysis. Although the response to the corresponding sine wave is rather smaller than that to the double impulse in the larger input level, it does not cause a serious problem. This is because the practical interest is up to about the input level $V/V_y = 3$.
5. The derived upper bound certainly bounds the actual plastic deformation after the second impulse in the first story, and the derived lower bound approximately circumvents the actual response from the lower side.
6. The actual critical response of plastic deformation after the second impulse in the first story is close to the upper or lower bound at

discrete input levels. Such actual critical response at the discrete level was analyzed by using the fundamental law in dynamics (Newton's second law).

7. The present method using the double impulse and the upper bound is applicable to actual recorded pulse-type ground motions within reasonable accuracy.

In this chapter, the case has been treated where the second story does not go into the plastic range after the first impulse. As stated before, this case is often the case when the condition, Eq. (10.17), is satisfied. The other case could be discussed in the future if necessary. As for damping, it is well recognized that the viscous damping is not effective for impulsive ground motions like near-fault ground motions. The effect in an elastic single-degree-of-freedom model is shown in Appendix 5. The effect of damping on the conservativeness of the proposed upper bound should be discussed in the future. The method may be applied to the analysis of base-isolated buildings or other-type flexible buildings subjected to near-fault ground motions (Hall et al. 1995, Iwan 1997, Vassiliou et al. 2013). Furthermore, its extension to the analysis of base-isolated buildings under long-duration, long-period ground motions may be possible (Saotome et al. 2018).

REFERENCES

Abbas, A. M., and Manohar, C. S. (2002). Investigations into critical earthquake load models within deterministic and probabilistic frameworks, *Earthquake Eng. Struct. Dyn.*, 31: 813–832.

Ben-Haim, Y., Chen, G., and Soong, T. T. (1996). Maximum structural response using convex models, *J. Engrg. Mech.*, ASCE, 122(4): 325–333.

Ben-Haim, Y., and Elishakoff, I. (1990). *Convex models of uncertainty in applied mechanics*, Elsevier.

Bertero, V. V., Mahin, S. A., and Herrera, R. A. (1978). Aseismic design implications of near-fault San Fernando earthquake records, *Earthquake Eng. Struct. Dyn.*, 6(1): 31–42.

Drenick, R. F. (1970). Model-free design of aseismic structures, *J. Eng. Mech. Div.*, ASCE, 96(EM4): 483–493.

Hall, J. F., Heaton, T. H., Halling, M. W., and Wald, D. J. (1995). Near-source ground motion and its effects on flexible buildings, *Earthquake Spectra*, 11(4): 569–605.

Iwan, W. D. (1997). Drift spectrum: Measure of demand for earthquake ground motions, *J. Structural Engineering*, ASCE, 123(4): 397–404.

Kojima, K., Fujita, K., and Takewaki, I. (2015). Critical double impulse input and bound of earthquake input energy to building structure, *Frontiers in Built Environment*, 1: 5.

Kojima, K., and Takewaki, I. (2015a). Critical earthquake response of elastic-plastic structures under near-fault ground motions (Part 1: Fling-step input), *Frontiers in Built Environment*, 1: 12.

Kojima, K., and Takewaki, I. (2015b). Critical earthquake response of elastic-plastic structures under near-fault ground motions (Part 2: Forward-directivity input), *Frontiers in Built Environment*, 1: 13.

Kojima, K., and Takewaki, I. (2015c). Critical input and response of elastic-plastic structures under long-duration earthquake ground motions, *Frontiers in Built Environment*, 1: 15.

Moustafa, A., Ueno, K., and Takewaki, I. (2010). Critical earthquake loads for SDOF inelastic structures considering evolution of seismic waves, *Earthquakes and Structures*; 1(2): 147–162.

Saotome, Y., Kojima, K., and Takewaki, I. (2018). Earthquake response of 2DOF elastic-perfectly plastic model under multiple impulse as substitute for long-duration earthquake ground motions, *Frontiers in Built Environment*, 4: 81.

Takewaki, I. (1996). Design-oriented approximate bound of inelastic responses of a structure under seismic loading, *Computers and Structures*, 61(3): 431–440.

Takewaki, I. (1997). Design-oriented ductility bound of a plane frame under seismic loading, *J. Vibration and Control*, 3(4): 411–434.

Takewaki, I. (2002). Robust building stiffness design for variable critical excitations, *J. Struct. Eng.*, ASCE, 128(12): 1565–1574.

Takewaki, I. (2007). *Critical excitation methods in earthquake engineering*, Oxford, Elsevier (Second edition in 2013).

Taniguchi, R., Kojima, K., and Takewaki, I. (2016). Critical response of 2DOF elastic-plastic building structures under double impulse as substitute of near-fault ground motion, *Frontiers in Built Environment*, 2: 2.

Vassiliou, M. F., Tsiavos, A., and Stojadinovic, B. (2013). Dynamics of inelastic base-isolated structures subjected to analytical pulse ground motions, *Earthquake Eng. Struct. Dyn.*, 42, 2043–2060.

APPENDIX 1: ADJUSTMENT OF AMPLITUDES OF DOUBLE IMPULSE AND CORRESPONDING ONE-CYCLE SINUSOIDAL WAVE

The adjustment of amplitudes of the double impulse and the corresponding one-cycle sinusoidal wave is achieved based on the equivalence of the maximum value of the Fourier amplitude (see Section 1.2 in Chapter 1). Figure10. A1 shows an example.

APPENDIX 2: UPPER AND LOWER BOUNDS OF PLASTIC DEFORMATION IN FIRST STORY AFTER SECOND IMPULSE (CASE OF ELASTIC RESPONSE IN FIRST STORY AFTER FIRST IMPULSE)

In the case of elastic response in the first story after the first impulse, another formulation is required. At the phase (D), the energy balance law can be expressed by

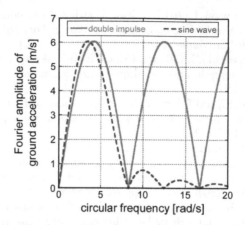

Figure 10.A1 Adjustment of amplitudes of double impulse and corresponding one-cycle sinusoidal wave based on Fourier amplitude equivalence (Taniguchi et al. 2016).

$$\frac{1}{2}(m_1 + m_2)V^2 + \Delta E = \frac{1}{2}k_1 d_{y1}^2 + k_1 d_{y1} d_{p1}^{(2)} + E_2^{(D)} + k_2 d_{y2} d_{p2} \quad (10.A1)$$

Eq. (10.A1) corresponds to Eq. (10.9) for the case of plastic response in the first story after the first impulse. It can be rearranged into

$$\frac{d_{p1}^{(2)}}{d_{y1}} = \frac{1}{k_1 d_{y1}^2}\left\{\frac{1}{2}(m_1 + m_2)V^2 + \Delta E - \left(E_2^{(D)} + k_2 d_{y2} d_{p2}^{(2)}\right)\right\} - \frac{1}{2} \quad (10.A2)$$

[Upper bound]

In Eq. (10.A2), the upper bound of plastic deformation $d_{p1}^{(2)}$ in the first story after the second impulse can be derived by maximizing the input energy ΔE by the second impulse and minimizing $E_2^{(D)}$ and $d_{p2}^{(2)}$. Then, the upper bound can be obtained from

$$\Delta E_{max} = \frac{3}{2}(m_1 + m_2)V^2 \quad (10.A3)$$

$$E_{2\,min}^{(D)} = 0 \quad (10.A4)$$

$$d_{p2\,min}^{(2)} \doteq 0 \quad (10.A5)$$

In this case, the upper bound of plastic deformation in the first story after the second impulse can be derived as

$$\frac{d_{p1}^{(2)}}{d_{y1}} \le \frac{1}{k_1 d_{y1}^2}\left\{2a^2\left(k_1 d_{y1}^2 + k_2 d_{y2}^2\right)\right\} - \frac{1}{2} \quad (10.A6)$$

For the model of equal mass, equal story stiffness and equal yield interstory drift, Eq. (10.A6) is reduced to

$$\frac{d_{p1}^{(2)}}{d_{y1}} \leq 4a^2 - \frac{1}{2} \tag{10.A7}$$

[Lower bound]

In Eq. (10.A2), the lower bound of plastic deformation $d_{p1}^{(2)}$ in the first story after the second impulse can be derived by minimizing ΔE and maximizing $E_2^{(D)}$ and $d_{p2}^{(2)}$. It is usual that the second story is in the elastic range even after the second impulse in a rather smaller input level. Then, the lower bound can be obtained from

$$\Delta E_{\min} = \frac{1}{2}(m_1 + m_2)V^2 + V\sqrt{m_1 k_1 d_{y1}^2} \tag{10.A8}$$

$$E_{2\max}^{(D)} \doteq \frac{1}{2} k_2 d_{y2}^2 \tag{10.A9}$$

$$d_{p2\max}^{(2)} \doteq 0 \tag{10.A10}$$

In this case, the lower bound of plastic deformation in the first story after the second impulse can be expressed as

$$\begin{aligned}
\frac{d_{p1}^{(2)}}{d_{y1}} &\geq \frac{1}{k_1 d_{y1}^2}\left\{a^2\left(k_1 d_{y1}^2 + k_2 d_{y2}^2\right)\right. \\
&\left. +a\sqrt{\frac{m_1}{m_1 + m_2}\left(k_1 d_{y1}^2 + k_2 d_{y2}^2\right)k_1 d_{y1}^2 - \frac{1}{2}k_2 d_{y2}^2}\right\} - \frac{1}{2}
\end{aligned} \tag{10.A11}$$

For the model of equal mass, equal story stiffness and equal yield interstory drift, Eq. (10.A11) is reduced to

$$\frac{d_{p1}^{(2)}}{d_{y1}} \geq 2a^2 + a - 1 \tag{10.A12}$$

APPENDIX 3: COMPARISON OF TIME HISTORIES TO DOUBLE IMPULSE AND CORRESPONDING SINUSOIDAL WAVE

Consider an undamped EPP 2DOF model of equal mass, equal story stiffness and equal yield interstory drift as treated above. Figures 10.A2 and 10.A3 show the comparison of the time histories (first-story interstory drift and

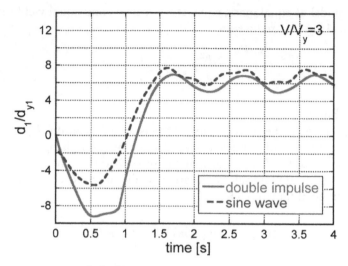

(a) First-story interstory drift

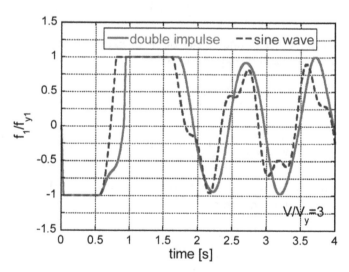

(b) First-story restoring force

Figure 10.A2 Comparison of time histories of first-story interstory drift and first-story restoring force under double impulse and corresponding sinusoidal wave: (a) first-story interstory drift, (b) first-story restoring force (Taniguchi et al. 2016).

first-story restoring force) and the first-story restoring-force characteristic, respectively, under the double impulse and the corresponding sinusoidal wave.

APPENDIX 4: INPUT LEVEL OF DOUBLE IMPULSE FOR CHARACTERIZING CRITICAL RESPONSE CLOSE TO UPPER OR LOWER BOUND

In Figure 10.12, the actual critical response of plastic deformation after the second impulse in the first story is close to the upper or lower bound at discrete input levels. The responses to the double impulse with such discrete input levels are investigated here.

Consider the deformation phase shown in Figure 10.15. At the phase (A), both masses are at rest and have zero velocities. In this case, the impulse (force × time interval) due to the restoring force in the first story is equal to the change of the sum of momenta at the input level $V = V_1$ (see Figure 10.A4).

$$(m_1 + m_2)V_1 = \int_0^{t_1} f_1(t)dt \qquad (10.A13)$$

where t_1 is the time at the phase (A). The restoring forces after the first impulse are treated as positive values in this Appendix. For another input level $V = V_1 + \Delta V$, a similar relation holds (see Figure 10.A4).

$$(m_1 + m_2)(V_1 + \Delta V) = \int_0^{t_1 + \Delta t} f_1(t)dt \qquad (10.A14)$$

where $t_1 + \Delta t$ is the time at the phase (A) for the input level $V = V_1 + \Delta V$. Since the time up to the first yielding in the first story is quite short, the following relation can be drawn approximately by subtracting Eq. (10.A13) from Eq. (10.A14).

$$(m_1 + m_2)\Delta V = \int_{t_1}^{t_1 + \Delta t} f_1(t)dt \qquad (10.A15)$$

Because $\Delta t = 2\pi/\omega$ is the period of the second story defined by Eq. (10.14a) with a circular frequency ω and $f_1 = k_1 d_{y1}$, the following relation holds.

$$(m_1 + m_2)\Delta V = 2\pi \sqrt{\frac{m_1 m_2}{(m_1 + m_2)k_2}} k_1 d_{y1} \qquad (10.A16)$$

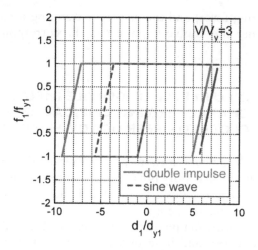

Figure 10.A3 Comparison of first-story restoring-fore characteristic under double impulse and corresponding sinusoidal wave (Taniguchi et al. 2016).

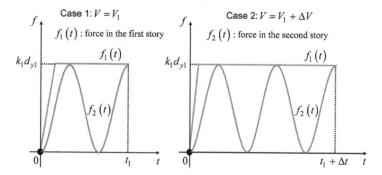

Figure 10.A4 Time histories of restoring forces in the first and second stories up to phase (A) (Taniguchi et al. 2016).

Normalizing Eq. (10.A16) by $V_y = \sqrt{\left(k_1 d_{y1}^2 + k_2 d_{y2}^2\right) / \left(m_1 + m_2\right)}$ defined in Eq. (10.3), the discrete input level ΔV is derived as

$$\frac{\Delta V}{V_y} = 2\pi \frac{\sqrt{m_1 m_2}}{m_1 + m_2} \frac{k_1 d_{y1}}{\sqrt{k_2 \left(k_1 d_{y1}^2 + k_2 d_{y2}^2\right)}} \qquad (10.A17)$$

In the undamped EPP 2DOF model of equal mass, equal story stiffness, and equal yield interstory drift as treated above, the following result is obtained.

$$\Delta V / V_y = 2.22 \qquad (10.A18)$$

This certainly corresponds to the discrete interval of input level observed in Figures 10.12 and 10.13.

APPENDIX 5: EFFECT OF VISCOUS DAMPING

To investigate the effect of viscous damping on the response under impulsive loading, consider an elastic damped SDOF model. Let ω_1, h denote the undamped natural circular frequency and the damping ratio. When the model is subjected to an initial velocity v_0 under the condition of 0 initial displacement, the displacement response can be expressed by

$$u = \frac{v_0}{\sqrt{1-h^2}\,\omega_1} e^{-h\omega_1 t} \sin\sqrt{1-h^2}\,\omega_1 t \qquad (10.A19)$$

When a one-cycle sinusoidal input corresponding to the double impulse is considered, the phase $\omega_1 t$ of the first peak corresponds to $\pi/2$ and the phase of the second peak corresponds to $3\pi/2$. In the case of the damping ratio $h = 0.02$, the ratio $e^{-h\omega_1 t}/\sqrt{1-h^2}$ of the displacement amplitude of the damped model to that of the undamped model is evaluated by

$$e^{-h\omega_1 t}/\sqrt{1-h^2} = 0.97 \text{ for } \omega_1 t = \pi/2 \qquad (10.A20a)$$

$$e^{-h\omega_1 t}/\sqrt{1-h^2} = 0.91 \text{ for } \omega_1 t = \pi/2 \qquad (10.A20b)$$

This indicates that the damping effect is only 3% for the first peak and 9% for the second peak. This indicates that the viscous damping effect may be small in case of impulsive loading.

Chapter 11

Optimal viscous damper placement for an elastic–perfectly plastic MDOF building model under critical double impulse

11.1 INTRODUCTION

Oil dampers are passive dampers of viscous type effective for broader amplitude ranges and the stiffness does not exist. This property is advantageous in the structural design of most framed building structures because the change of stiffness in building structures due to the use of passive dampers usually leads to the variation of design loads and the necessity of change of member size. Various theories of optimal damper placement have been proposed (see Takewaki 2009, Domenico et al. 2019). Although tuned-mass dampers were only passive dampers around 1970s and 1980s, interstory-type passive dampers were introduced gradually around 1990.

In an early stage, most researches on damper optimization (quantity and location in interstory-type passive dampers) are limited to elastic problems. The work by Zhang and Soong (1992) may be a milestone in this direction. They proposed a simple algorithm to insert dampers sequentially into the location exhibiting the maximum response. Tsuji and Nakamura (1996) presented an optimality condition-based sequential algorithm for optimal damper placement. The work by Takewaki (1997) became another milestone in the field of damper optimization. He proposed an incremental inverse problem approach to the investigation of optimal damper placement by using the transfer function amplitude (approximately equivalent to the top displacement in the lowest mode) as an objective performance. Soong introduced this approach in the second world conference on structural control in Kyoto (Soong 1998). As another approach, Takewaki et al. (1999) introduced a sensitivity-based approach in the field of optimal damper placement.

In the 21st century, Garcia (2001) extended the approach by Zhang and Soong (1992) and compared the optimization performances of several algorithms including the method by Takewaki (1997). Singh and Moreschi (2001) investigated optimal design problems using the optimality conditions and the nonlinear programming. Uetani et al. (2003) presented a practical and general damper optimization method for nonlinear framed structures based on the mathematical programming and applied its method to an actual high-rise building in Osaka. Lavan and Levy (2006) introduced a methodology for the optimal design of added viscous damping for an ensemble of realistic ground motion records with a constraint on the maximum drift. They transformed the original problem into some equivalent problems. Silvestri and Trombetti (2007) proposed physical and numerical approaches for the optimal insertion of seismic viscous dampers in shear-type structures and compared the optimization performances of several algorithms including the method by Takewaki (1997). Aydin et al. (2007) investigated the optimal damper distribution for seismic rehabilitation of planar building structures. Whittle et al. (2012) compared some viscous damper placement methods including the method by Takewaki (1997) for improving seismic building design.

As far as nonlinear dampers or nonlinear structures with dampers are concerned, a limited number of investigations has been proposed. Lavan and Levy (2005) investigated a problem of optimal design of supplemental viscous dampers for yielding irregular shear-frames. Attard (2007) treated a problem of optimal viscous damping for controlling interstory displacements in highly nonlinear steel buildings. Lavan et al. (2008) presented a noniterative optimization procedure for seismic weakening and damping of inelastic structures. Adachi et al. (2013) proposed a practical method of optimal relief-force distribution (for mitigating the failure of dampers) for oil dampers by setting the maximum interstory drift and the maximum building-top acceleration as the objective performances. Murakami et al. (2013) tackled a problem of simultaneous optimal damper placement using oil, hysteretic, and inertial mass dampers and proposed a sensitivity-based algorithm. Since different-type dampers possess different-type properties, the cost ratio among oil, hysteretic, and inertial mass dampers may be a key point in such problem. In addition, the irregular unstable sensitivity for hysteretic dampers and recorded earthquake ground motions seems to be a difficult but challenging issue. Pollini et al. (2017) dealt with a problem of optimal placement of nonlinear viscous dampers by using the adjoint sensitivity analysis method. Shiomi et al. (2018) investigated a problem of optimal hysteretic damper placement for elastic-plastic multi-degree-of-freedom (MDOF) shear building models under the double impulse as a representative version of near-fault ground motions and proposed a sensitivity-based method. Their approach defines the shear building model with uniform damper distribution as an initial design and reduces the unnecessary dampers based on the sensitivity information. Recently, Domenico et al. (2019) presented a well-documented review on the optimization of passive dampers and provided a new perspective for future research.

However, there is no method enabling an efficient analysis of optimal viscous damper placement for MDOF building structures that experience rather large plastic deformation.

In this chapter, a new method for optimal viscous damper placement is explained for elastic–perfectly plastic (EPP) MDOF shear building structures subjected to the critical double impulse as a representative version of near-fault ground motions (Akehashi and Takewaki 2019). The critical interval between two impulses of the double impulse is characterized by the criterion on the maximum input energy by the second impulse. The critical timing of the second impulse is proved to be the timing at which the sum of the restoring force and the damping force in the first story attains zero in the unloading process after the first impulse. The objective function and constraint in terms of the maximum interstory drift along the building height or the sum of the maximum interstory drifts in all stories are chosen and the corresponding optimization algorithm based on time-history response analysis and sensitivity analysis is presented. Since the objective function in terms of the sum of the maximum interstory drifts in all stories is superior to the objective function in terms of the maximum interstory drift along the building height, it is employed in this chapter. In addition, a new concept called "the double impulse pushover (DIP)," an extended version of incremental dynamic analysis (IDA) by Vamvatsikos and Cornell (2001), is introduced and explained for determining the input velocity level of the critical double impulse. DIP indicates the dynamic pushover that seeks for only the critical resonant case. The combination of two algorithms is shown to be promising and effective for finding a stable optimal damper placement, i.e., one for effective reduction of the overflowed maximum interstory drift via the concentrated allocation of dampers and the other for effective allocation of dampers via the use of a stable objective function.

11.2 INPUT GROUND MOTION

Earthquake ground motions are intrinsically uncertain in occurrence and properties (for example, Abrahamson et al. 1998). It is also recognized that near-fault ground motions possess peculiar characteristics, e.g., pulse-type waves (Bertero et al. 1978, Mavroeidis and Papageorgiou 2003, Kalkan and Kunnath 2006). To capture such peculiar characteristics of near-fault ground motions, Kojima and Takewaki (2015) introduced the double impulse as a simplified representative version of the pulse-shaped main portion of near-fault ground motions. The acceleration $\ddot{u}_g(t)$ of the double impulse with the time interval t_0 of two impulses can be expressed by

$$\ddot{u}_g(t) = V\delta(t) - V\delta(t - t_0),\qquad(11.1)$$

where V is the input velocity amplitude and $\delta(t)$ is the Dirac delta function. Figure 11.1(a) shows the acceleration, velocity, and displacement of the double impulse together with those of the corresponding one-cycle sine wave of

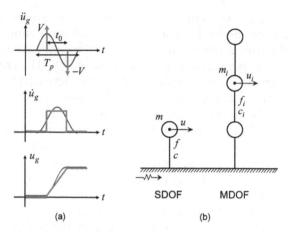

Figure 11.1　Double impulse and structural models: (a) acceleration, velocity, and displacement of double impulse and corresponding one-cycle sine wave: (b) SDOF model and MDOF model (Akehashi and Takewaki 2019).

acceleration. In the response evaluation stage, it is important to adjust the input level of the double impulse and the corresponding one-cycle sinusoidal wave based on the equivalence of the maximum Fourier amplitude as noted in Section 1.2 in Chapter 1.

While several investigations using the double impulse were made for single-degree-of-freedom (SDOF) models (Kojima and Takewaki 2015, 2016, Kojima et al. 2017, Akehashi et al. 2018a, b), researches on MDOF models are few (Taniguchi et al. 2016 and Chapter 10 in this book, Shiomi et al. 2018, Saotome et al. 2018). This is due to the fact that the simple energy balance law for the simple derivation of the maximum response is difficult to apply for MDOF models because of the phase lag among masses. Taniguchi et al. (2016) dealt with an undamped EPP 2DOF model under the double impulse and found, by using the criterion of the maximum input energy, that the timing of the second impulse at the zero restoring-force state becomes the critical timing. However, this critical timing cannot be used directly for damped models.

Consider an EPP SDOF model with viscous damping and an EPP multi-degree-of-freedom (MDOF) model with viscous damping as shown in Figure 11.1(b). Let us consider first the EPP SDOF model with viscous damping subjected to the double impulse. The critical input timing of the second impulse can be defined as the timing that maximizes the input energy to this model by the second impulse. To demonstrate this fact, consider the following equation of motion.

$$m\ddot{u} + c\dot{u} + f\left(k, u, d_y, d_r\right) = 0, \qquad (11.2)$$

where m, c, f, k, u, d_y, d_r denote the mass, damping coefficient, restoring force, initial stiffness, displacement, yield displacement, and residual displacement. As in other chapters, the super dot indicates the differentiation with respect to time. Assume that this EPP SDOF model with viscous damping is subjected to the double impulse and free vibration occurs. Since the displacement does not change instantaneously at the second impulse, the strain energy does not change at the time of action of the second impulse. Therefore, the input energy by the second impulse can be expressed by

$$E = \frac{1}{2}m\left(\dot{u} + V\right)^2 - \frac{1}{2}m\dot{u}^2 = \left(m\dot{u}\right)V + \frac{1}{2}mV^2. \quad (11.3)$$

In Eq. (11.3), $\dot{u}$ is the velocity just before the input of the second impulse. Eq. (11.3) indicates that when the velocity attains the maximum, the input energy by the second impulse yields the maximum. The condition that $\dot{u}$ attains the extremum is $\ddot{u} = 0$. Substitution of $\ddot{u} = 0$ into Eq. (11.2) leads to $c\dot{u} + f = 0$. This means that when the sum of the restoring force and the damping force becomes zero (i.e., the velocity becomes the maximum), the input energy by the second impulse becomes the maximum.

Consider an N-story EPP MDOF shear building model with viscous damping in the next step. The mass in the i-th story is denoted by m_i. Let u_i denote the horizontal displacement of mass m_i. As in the EPP SDOF model with viscous damping, since the displacements do not change instantaneously at the action of the second impulse, the strain energy does not change at the time of action of the second impulse. Therefore, the input energy by the second impulse can be expressed by

$$E = \sum_{i=1}^{N}\frac{1}{2}m_i\left(\dot{u}_i + V\right)^2 - \sum_{i=1}^{N}\frac{1}{2}m_i\dot{u}_i^2 = V\sum_{i=1}^{N}m_i\dot{u}_i + \sum_{i=1}^{N}\frac{1}{2}m_iV^2. \quad (11.4)$$

Eq. (11.4) means that when $\sum_{i=1}^{N}\left(m_i\dot{u}_i\right)$ attains the maximum, the input energy by the second impulse becomes the maximum. The condition that $\sum_{i=1}^{N}\left(m_i\dot{u}_i\right)$ attains the extremum is expressed by $\sum_{i=1}^{N}\left(m_i\ddot{u}_i\right) = 0$. Since $\sum_{i=1}^{N}\left(m_i\ddot{u}_i\right)$ is equal to $F_1 = c_1\dot{u}_1 + f_1$ owing to the dynamic equilibrium, i.e., the sum of the damping force and the restoring force in the first story, the extremum condition becomes $F_1 = c_1\dot{u}_1 + f_1 = 0$. This critical condition is very simple and can be used in the time-history response analysis in a simple manner.

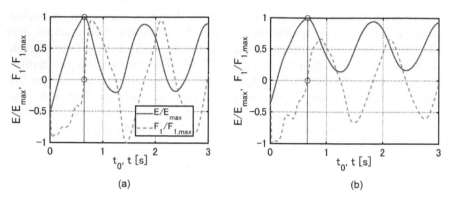

Figure 11.2 Examples of time history of $F_1 = c_1\dot{u}_1 + f_1$ for model subjected to double impulse including second impulse at time satisfying $F_1 = c_1\dot{u}_1 + f_1 = 0$ and variation of input energy E by second impulse with respect to t_0: (a) $V = 0.84$ [m/s], (b) $V = 1.23$ [m/s] (Akehashi and Takewaki 2019).

Figure 11.2 shows an example of the time history of $F_1 = c_1\dot{u}_1 + f_1$ for the EPP MDOF model with viscous damping subjected to the double impulse including the second impulse at the time satisfying $F_1 = c_1\dot{u}_1 + f_1 = 0$ and the variation of the input energy E by the second impulse with respect to t_0. Figure 11.3 presents the condition characterizing the critical timing of the second impulse in the double impulse.

The correspondence of the double impulse with recorded ground motions was investigated in the reference(supplementary material in Akehashi and Takewaki 2019).

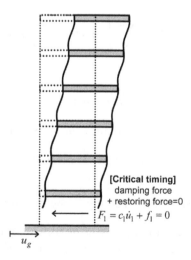

Figure 11.3 Condition characterizing critical timing of second impulse in double impulse.

11.3 PROBLEM OF OPTIMAL DAMPER PLACEMENT AND SOLUTION ALGORITHM

The direct problem of optimal damper placement may be stated as follows: to minimize the maximum interstory drift (or the sum of the maximum interstory drifts along height) under the condition on the specified total quantity of damping coefficients of passive dampers. Another problem may be described as: to minimize the total quantity of passive dampers under the constraint on the maximum interstory drift. These two problems may be proven to be almost equivalent as in the dual problem of mathematical optimization. It seems possible to deal with these problems by using the sensitivity-based approach that includes the time-history response analysis for the double impulse and the finite difference method for numerical sensitivity analysis. However, it was found that the direct application of this approach to the above-mentioned problems leads to unstable results. To overcome this difficulty, a mixed-type approach including the following two problems (Problem 1 and 2) was proposed in the paper (Akehashi and Takewaki 2019). The problems treated there will be explained next.

Consider first the following problem.

[Problem 1]

min $\mathbf{c}^T_{\text{add}} \cdot \mathbf{1}$ subject to: $d_{\text{max}, i} \le d_{\text{target}, i}$, for all i

In Problems 1, $\mathbf{c}_{\text{add}}$ is the damping coefficient vector for added dampers in all stories and $\mathbf{1}$ is the vector including unity in all components. The superscript T indicates the transpose. The quantity $d_{\text{max}, i}$ is the maximum interstory drift in the i-th story under the critical double impulse and $d_{\text{target}, i}$ is the target value of $d_{\text{max}, i}$. The solution algorithm for this problem may be described as follows.

[Algorithm 1]

Step 1 Input the critical double impulse to the bare MDOF model without dampers and compute the maximum interstory drifts. Put $j \to 0$.

Step 2 Investigate the stories i that satisfy $d^j_{\text{max},i} > d_{\text{target},i}$. Add the small damping coefficient Δc only in the i-th story. Input the critical double impulse to the modified MDOF model and compute the maximum interstory drifts. Evaluate the numerical sensitivity $s^j_i = d^j_{\text{max},i} - d^{j+1}_{\text{max},i}$ by the finite difference method.

Step 3 Find the largest value of s^j_i and update the damping coefficient as $c_i \to c_i + \Delta c$. If the model satisfies $d_{\text{max}, i} \le d_{\text{target}, i}$ for all i, then finalize the process. If not, put $j \to j+1$ and return to Step 2.

The flow of Algorithm 1 is shown in Figure 11.4(a). Algorithm 1 is intended to implement the effective reduction of the overflowed maximum

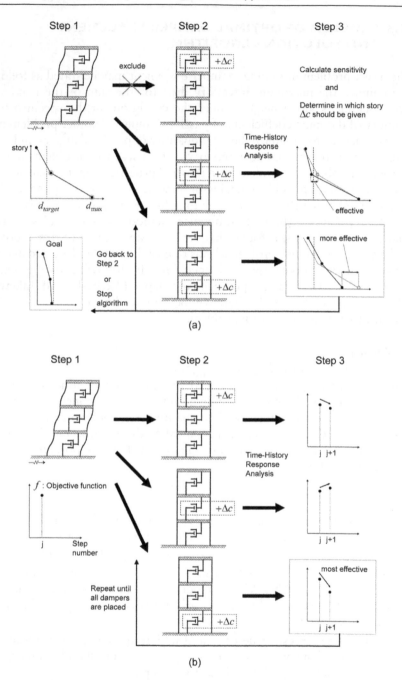

Figure 11.4 Overview of algorithms of optimal damper placement: (a) Algorithm 1, (b) Algorithm 2 (Akehashi and Takewaki 2019).

interstory drift (the stories i in Step 2) via the concentrated allocation of dampers. It was found after some attempts that Algorithm 1 sometimes encounters difficulties, i.e., inability to deal with the problem for a pre-scribed damper quantity.

Consider secondly the following problem.

[Problem 2]

$$\min f = \sum_{i=1}^{N} d_{\max,i} \text{ subject to: } \mathbf{c}^{T}_{\text{add}} \cdot \mathbf{1} = const.$$

The solution algorithm for this problem may be described as follows.

[Algorithm 2]

Step 1 Input the critical double impulse to the bare MDOF model without dampers and compute the objective function f. Put $j \to 0$.

Step 2 Make N models in each of which Δc is added in one of the first through N-th stories. Input the critical double impulse to each of those MDOF models and compute the objective function f.

Step 3 Find the story in which the largest reduction of f occurs. For that story i, update the damping coefficient as $c_i \to c_i + \Delta c$. Update the index $j \to j+1$. If $\Delta c \cdot j = \mathbf{c}^{T}_{\text{add}} \cdot \mathbf{1}$ is satisfied, then finalize the process. If $\Delta c \cdot j < \mathbf{c}^{T}_{\text{add}} \cdot \mathbf{1}$ is satisfied, return to Step 2.

The flow of Algorithm 2 is shown in Figure 11.4(b). It was found after some attempts that Algorithm 2 sometimes encounters difficulties in effi-ciency depending on models. The detailed explanation will be shown in numerical examples afterwards.

Consider thirdly the following mixed problem.

[Problem 3: Mixed Problem of Problems 1 & 2]

Problem 1 is solved at first until some stage and then Problem 2 is solved. The solution algorithm for this problem may be described as follows.

[Algorithm 3: Combination of Algorithms 1 & 2]

First of all, apply Algorithm 1. Set $d_{\text{target}, i} = d_{y, i}$ for all i and obtain the model in which the largest interstory drift attains the elastic limit. Adopt this model as another initial model and repeat Algorithm 2 in $\mathbf{c}^{T}_{\text{add}} \cdot 1/\Delta c$ steps.

Algorithm 3 is superior to Algorithm 1 and 2 because the combination of two algorithms, one for effective reduction of the overflowed maximum interstory drift via the concentrated allocation of dampers and the other for effective allocation of dampers via the use of the stable objective function, is effective for finding a stable optimal damper placement.

11.4 THREE MODELS FOR NUMERICAL EXAMPLES

Consider three shear building models of 12 stories with different story stiffness distributions. Model 1 has a uniform distribution of story stiffnesses. Model 2 has a straight-line lowest eigenmode (the story stiffnesses are obtained by the inverse-mode formulation). Model 3 has a stepped distribution of story stiffnesses (upper four stories, middle four stories, and lower four stories have uniform stiffness distributions with different values; the ratios among them are 1: 1.5: 2). The undamped fundamental natural period of these three models is 1.2(s) and the structural damping ratio is 0.01 (stiffness proportional type). All the floor masses have the same value. The common story height is 4 m and the common yield interstory drift ratio is 1/150. The story shear-interstory drift relation obeys the elastic–perfectly plastic (EPP) rule.

Figure 11.5 shows the eigenmodes multiplied by the participation factor (participation vectors) and the natural periods for three models.

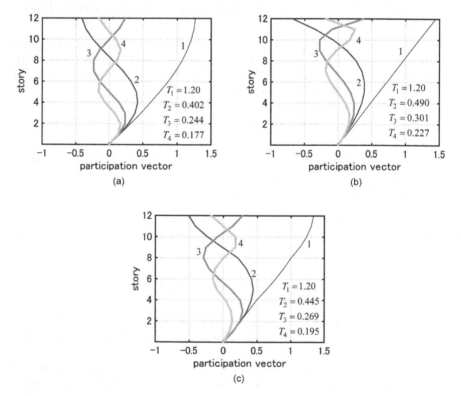

Figure 11.5 Eigenmodes multiplied by participation factors (participation vectors) and natural periods for three models: (a) Model 1, (b) Model 2, (c) Model 3 (Akehashi and Takewaki 2019).

11.5 DYNAMIC PUSHOVER ANALYSIS FOR INCREASING CRITICAL DOUBLE IMPULSE (DIP: DOUBLE IMPULSE PUSHOVER)

The Kumamoto earthquake (Japan) in 2016 strongly influenced the definition of design earthquake ground motions. The maximum ground velocity over 2 m/s is far from common sense in the earthquake-resistant design of structures. To determine the input velocity level of the critical double impulse, an idea similar to the incremental dynamic analysis (IDA) procedure (Vamvatsikos and Cornell 2001) is applied to the critical double impulse. It should be reminded that only the critical set (amplitude and time interval of the double impulse) is treated here. In other words, the interval of two impulses of the double impulse is varied depending on the input velocity level (also depending on the maximum interstory drift). Akehashi and Takewaki (2019) called this procedure "Double impulse pushover (DIP)." DIP provides the relation between the maximum interstory drift and the input velocity level of the critical double impulse. While the conventional IDA includes multiple recorded ground motions for taking into account the uncertainty in predominant periods of ground motions, DIP adopts the critical input and enables an efficient analysis of the relation between the maximum response and the input level.

Figure 11.6 shows the maximum interstory drift distributions by DIP. The velocity level is increased from $V = 0.2$ m/s to $V = 1.6$ m/s by 0.2 m/s.

Since the input velocity level of the critical double impulse influences the maximum interstory drift and the optimal damper placement greatly, its determination appears very important. The determination process of the input velocity level of the critical double impulse is explained in the next.

1. Specify the maximum interstory drift of the initial design model (bare model without dampers).
2. Conduct DIP for the initial design model. Find the velocity level V for which the maximum interstory drift of the initial design model exceeds the specified value for the first time. Conduct DIP also for larger values of the velocity level V.
3. Draw the maximum interstory drift distributions by DIP as shown in Figure 11.6 with respect to the input velocity level. Realize how easily the plastic deformation is concentrated to a special location. Based on these results, determine the input velocity level of the critical double impulse so that the maximum interstory drift exceeds the specified value.

Model 3 is treated as an example of determining the input level. The double of the yield interstory drift is taken as the specified target value of the maximum interstory drift. Then, it was found that over 0.6 m/s is necessary.

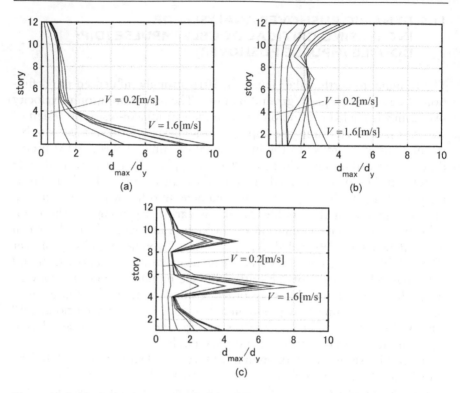

Figure 11.6 Maximum interstory drift by DIP: (a) Model 1, (b) Model 2, (c) Model 3 (Akehashi and Takewaki 2019).

Although the input velocity level should be chosen for each structural model, $V = 0.84$ m/s and 1.23 m/s are employed in the following section.

11.6 NUMERICAL EXAMPLES

11.6.1 Examples for Problem 1 using Algorithm 1

Consider first some examples for Problem 1. The amplitudes of the critical double impulses were determined from the results for the DIP analysis explained in Section 11.5.

Figure 11.7 shows the distribution of damping coefficients of added dampers, $\max(d_{\max, i}/d_y)$ with respect to step number and the distribution of $d_{\max, i}/d_y$ under the critical double impulses with $V = 0.84$ [m/s] and $V = 1.23$ [m/s] for Model 1. The condition $d_{\text{target}, i} = d_y$ (for all i) is adopted.

Figure 11.8 presents similar figures for Model 2 and Figure 11.9 illustrates those for Model 3.

Figures 11.7, 11.8, and 11.9 show that as the input velocity level increases, the ratios of damping coefficients of added dampers along height change

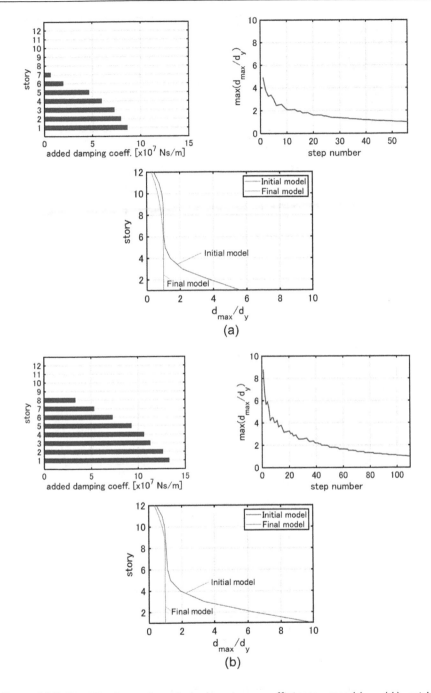

Figure 11.7 Distribution of added damping coefficients, $\max(d_{max,\,i}/d_y)$ with respect to step number and distribution of $d_{max,\,i}/d_y$ under critical double impulse for Model 1 (Problem 1): (a) $V = 0.84$ [m/s], (b) $V = 1.23$ [m/s] (Akehashi and Takewaki 2019).

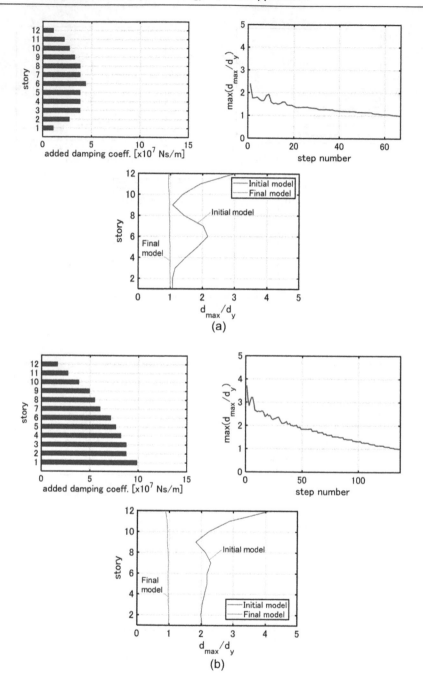

Figure 11.8 Distribution of added damping coefficients, max($d_{max, i}/d_y$) with respect to step number and distribution of $B_{max, i}/d_y$ under critical double impulse for Model 2 (Problem 1): (a) V = 0.84 [m/s], (b) V = 1.23 [m/s] (Akehashi and Takewaki 2019).

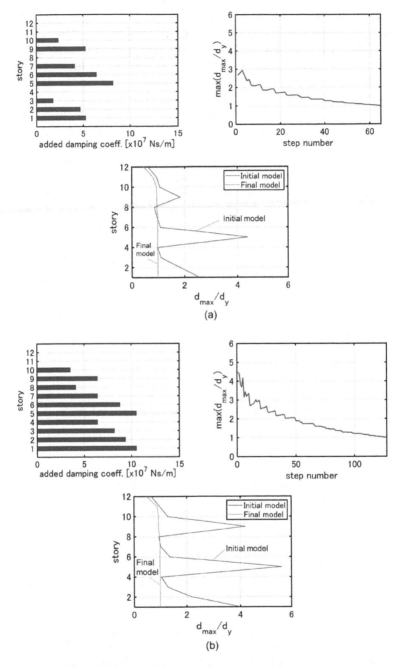

Figure 11.9 Distribution of added damping coefficients, $\max(d_{max, i}/d_y)$ with respect to step number and distribution of $d_{max, i}/d_y$ under critical double impulse for Model 3 (Problem 1): (a) $V = 0.84$ [m/s], (b) $V = 1.23$ [m/s]: (a) $V = 0.84$ [m/s], (b) $V = 1.23$ [m/s] (Akehashi and Takewaki 2019).

and their allocation becomes smooth in the wide height range. This is because as the input level becomes larger, the number of stories experiencing plastic deformation increases. Secondly, the maximum interstory drift distribution in the final model is controlled to become almost uniform by Algorithm 1. Furthermore, the increase of $\max(d_{max,i}/d_y)$ in the damper allocation process is allowed.

For Model 3, it can be seen that the dampers are not allocated in the fourth and eighth stories for the input level $V = 0.84$ [m/s], but those are allocated for the input level $V = 1.23$ [m/s]. However, the maximum interstory drifts in those stories are smaller than the elastic limit in the initial stage. This indicates that Algorithm 1 is applied first so that the dampers are allocated to the first, fifth, and ninth stories experiencing large plastic deformation. This process helps the energy required for inducing deformation distribute to the neighboring stories. As a result, among the stories neighboring to the fifth and ninth stories, the deformations in the sixth and tenth stories with relatively small stiffness becomes larger. Since the fourth and eighth stories with relatively large stiffness go into the plastic range for the input level of $V = 1.23$ [m/s], the dampers are allocated so as to strengthen the model. A similar observation may be possible also for Model 1 and 2.

It can be summarized that Algorithm 1 is more apt to allocate added dampers to the stories where the plastic deformation develops first then to allocate additional ones to rather weak stories after the strengthening is completed.

11.6.2 Examples for Problem 2 using Algorithm 2

Consider some examples for Problem 2 in the following. The parameter specification $100\Delta c = \mathbf{c}^T_{add} \cdot \mathbf{1}$ is given here.

Figure 11.10 shows the distribution of damping coefficients of added dampers, $\sum d_{max,i}/d_y$ with respect to step number and the distribution of $d_{max,i}/d_y$ under the critical double impulses with $V = 0.84$ [m/s] and $V = 1.23$ [m/s] for Model 1. The distributions for the elastic limit are also shown for reference $(\max(d_{max,i}/d_y) = 1)$.

Figure 11.11 presents similar figures for Model 2 and Figure 11.12 illustrates those for Model 3.

It can be observed that when the elastic limit is employed as the target for determining the input velocity level, the dampers are allocated so that the maximum interstory drifts become almost uniform for all models (Models 1–3). For Model 1, the dampers are allocated so that the maximum interstory drifts become almost uniform regardless of the input velocity level. On the other hand, for Models 2 and 3, the damper distributions are different depending on the input velocity level. Furthermore, for Models 2 and 3, the maximum interstory drifts in specific stories do not change between the initial model and the final model. This means that, since Algorithm 2 is aimed at finding the optimal damper allocation by seeking the steepest

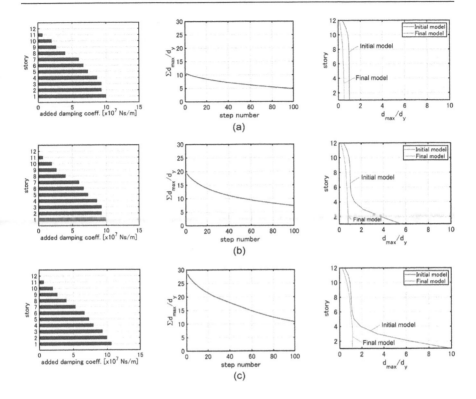

Figure 11.10 Distribution of added damping coefficients, $\sum d_{max,i}/d_y$ with respect to step number and distribution of $d_{max,i}/d_y$ under critical double impulse for Model 1 (Problem 2): (a) elastic limit, (b) $V = 0.84$ [m/s], (c) $V = 1.23$ [m/s] (Akehashi and Takewaki 2019).

direction (most effective direction for deformation reduction), it does not provide better damper allocation from the viewpoint of uniform reduction of the maximum interstory drifts in all stories.

11.6.3 Examples for Mixed Problem (Problem 3) of Problem 1 and 2 using Algorithm 3

Examples for Mixed Problem (Problem 3) of Problem 1 and 2 using Algorithm 3 are presented in this section. The parameter specification by $100\Delta c = \mathbf{c}^T_{add} \cdot \mathbf{1}$ (100 step allocation of total dampers) is used for $V = 0.84$ [m/s] and the parameter specification by $250\Delta c = \mathbf{c}^T_{add} \cdot \mathbf{1}$ (250 step allocation of total dampers) is used for $V = 1.23$ [m/s].

Figure 11.13 shows the distribution of damping coefficients of added dampers, $\sum d_{max,i}/d_y$ with respect to step number and the distribution of $d_{max,i}/d_y$ under the critical double impulses with $V = 0.84$ [m/s] and $V = 1.23$ [m/s] for Model 1.

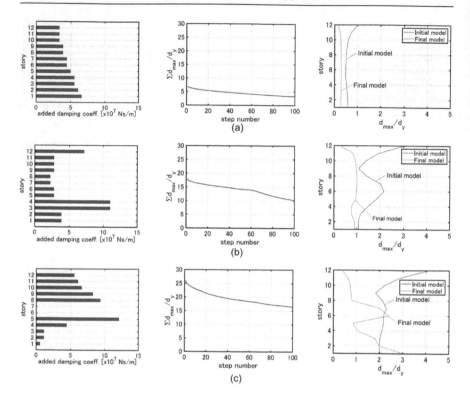

Figure 11.11 Distribution of added damping coefficients, $\sum d_{max,i}/d_y$ with respect to step number and distribution of $d_{max,i}/d_y$ under critical double impulse for Model 2 (Problem 2): (a) elastic limit, (b) $V = 0.84$ [m/s], (c) $V = 1.23$ [m/s] (Akehashi and Takewaki 2019).

Figure 11.14 presents similar figures for Model 2 and Figure 11.15 illustrates those for Model 3.

From the above analysis, we can observe the following facts. The analysis in the elastic range is not easy by Algorithm 1 for Problem 1 owing to the inability to set the total damper quantity and the maximum interstory drift distribution of the final model obtained by Algorithm 2 for Problem 2 is unstable (not uniform) depending on the model. On the other hand, a stable damper allocation is possible by Algorithm 3 for Problem 3 regardless of the input velocity level of the critical double impulse and the final maximum interstory drift distributions are apt to become uniform. From these results, it may be said that Algorithm 3 enables the procedure to guarantee the minimum performance by using a small amount of added dampers and to reduce the structural response in a global sense by using the additional amount of added dampers.

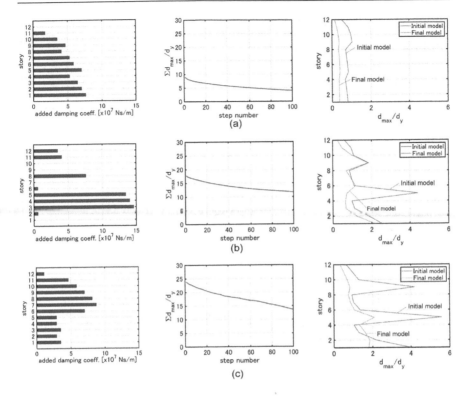

Figure 11.12 Distribution of added damping coefficients, $\sum d_{max,\,i}/d_y$ with respect to step number and distribution of $d_{max,\,i}/d_y$ under critical double impulse for Model 3 (Problem 2): (a) elastic limit, (b) $V = 0.84$ [m/s], (c) $V = 1.23$ [m/s] (Akehashi and Takewaki 2019).

11.7 COMPARISON OF IDA (INCREMENTAL DYNAMIC ANALYSIS) AND DIP

Akehashi and Takewaki (2019) extended the critical excitation method for SDOF elastic-plastic systems to that for MDOF elastic-plastic systems. They characterized the criticality of the double impulse input as the maximum input energy by the second impulse (see Figure 11.3). Furthermore, Akehashi and Takewaki (2019) introduced a new concept of "double impulse push-over" to make the inelastic response property clearer. Figure 11.2 shows the time-history of the base shear $F_1 = c_1 \dot{u}_1 + f_1$ of an MDOF elastic-plastic shear building model and the input energy E by the second impulse with respect to the impulse interval t_0. It can be understood that the timing of the second impulse at the zero base shear ($F_1 = c_1 \dot{u}_1 + f_1 = 0$) gives the critical timing of the second impulse which leads to the maximum input energy. To capture the elastic-plastic response characteristics in a clearer way, incremental

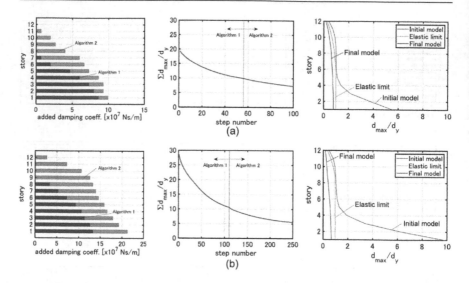

Figure 11.13 Distribution of added damping coefficients, $\sum d_{max,\,i}/d_y$ with respect to step number and distribution of $d_{max,\,i}/d_y$ under critical double impulse for Model 1 (Problem 3): (a) $V = 0.84$ [m/s], (b) $V = 1.23$ [m/s] (Akehashi and Takewaki 2019).

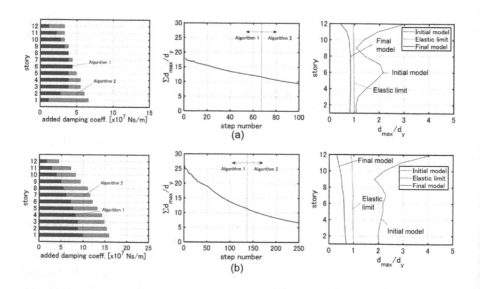

Figure 11.14 Distribution of added damping coefficients, $\sum d_{max,\,i}/d_y$ with respect to step number and distribution of $d_{max,\,i}/d_y$ under critical double impulse for Model 2 (Problem 3): (a) $V = 0.84$ [m/s], (b) $V = 1.23$ [m/s] (Akehashi and Takewaki 2019).

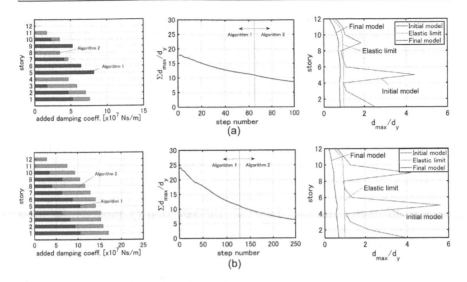

Figure 11.15 Distribution of added damping coefficients, $\sum d_{max,i}/d_y$ with respect to step number and distribution of $d_{max,i}/d_y$ under critical double impulse for Model 3 (Problem 3): (a) $V = 0.84$ [m/s], (b) $V = 1.23$ [m/s] (Akehashi and Takewaki 2019).

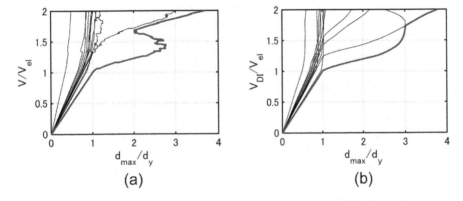

Figure 11.16 Demonstration of elastic-plastic response characteristics: (a) incremental dynamic analysis (IDA) due to Vamvatsikos and Cornell (2001); (b) double impulse pushover (DIP) due to Akehashi and Takewaki (2020) (each line shows trajectory of maximum inter-story drift in each story with respect to input velocity level) (Courtesy of Mr. H. Akehashi of Kyoto University).

analysis with respect to input motion level may be desirable. Figure 11.16 presents the comparison between the incremental dynamic analysis (IDA) due to Vamvatsikos and Cornell (2001) (this figure was plotted only for El Centro NS 1940) and the double impulse pushover (DIP) due to Akehashi and Takewaki (2020). Each line shows the trajectory of the maximum interstory drift in each story with respect to input velocity level. The envelope line is also shown in Figure 11.16. It should be remarked that, while IDA does not provide the critical case for each stage of the input ground motion level, DIP gives the critical case for each input ground motion level.

11.8 SUMMARIES

An innovative method for optimal viscous damper placement was explained for elastic–perfectly plastic (EPP) multi-degree-of-freedom (MDOF) shear building structures subjected to the critical double impulse as a representative of near-fault ground motions. Simultaneous treatment of nonlinear MDOF structures and uncertainty in the selection of input is the most remarkable point that has never been overcome in the past. The main results can be summarized as follows.

1. The method explained in this chapter consists of two phases: i) rapid reduction of the overflowed maximum inelastic interstory drift by the effective insertion of viscous dampers to concentrated stories; and ii) effective damper allocation using a stable sensitivity analysis of the objective function. The sensitivity analysis employs the time-history response analysis for the critical double impulse and the finite difference analysis for sensitivity analysis of damping coefficients.
2. The adoption of the critical timing of the second impulse based on the criterion on the maximum input energy enables an efficient analysis of optimal damper placement. This criterion tremendously reduces the analysis load for finding the critical timing. This process usually requires repetition and excessive computational load in conventional methods. This criterion on the critical timing of the second impulse is general and is applicable to any elastic-plastic MDOF structures.
3. A new concept of double impulse pushover (DIP) was introduced and explained for determining the appropriate input velocity level of the critical double impulse. DIP can be regarded as a superior dynamic pushover analysis tool.
4. The critical double impulse enables an efficient analysis of optimal damper placement for near-fault ground motions.
5. The explained method is useful for broad-type building structures with various stiffness distributions.

REFERENCES

Abrahamson, N., Ashford, S., Elgamal, A., Kramer, S., Seible, F., and Somerville, P. (1998). 1st PEER Workshop on Characterization of Special Source Effects, Pacific Earthquake Engineering Research Center, University of California, San Diego, 1998.

Akehashi, H., Kojima, K., and Takewaki, I. (2018a). Critical response of SDOF damped bilinear hysteretic system under double impulse as substitute for near-fault ground motion, *Frontiers in Built Environment*, 4: 5.

Akehashi, H., Kojima, K., Fujita, K., and Takewaki, I. (2018b). Critical Response of Nonlinear Base-Isolated Building Considering Soil-Structure interaction under double impulse as substitute for near-fault ground motion, *Frontiers in Built Environment*, 4: 34.

Akehashi, H., and Takewaki, I. (2019). Optimal viscous damper placement for elastic-plastic MDOF structures under critical double impulse, *Frontiers in Built Environment*, 5: 20.

Attard, T. L. (2007). Controlling all interstory displacements in highly nonlinear steel buildings using optimal viscous damping, *J Struct Eng*, ASCE, 133(9): 1331–1340.

Aydin, E., Boduroglub, M. H., and Guney, D. (2007). Optimal damper distribution for seismic rehabilitation of planar building structures, *Eng Struct*, 29: 176–185.

Bertero, V. V., Mahin, S. A., and Herrera, R. A. (1978). Aseismic design implications of near-fault San Fernando earthquake records, *Earthquake Eng Struct Dyn*. 6(1): 31–42.

Domenico, D. D., Ricciardi, G., and Takewaki, I. (2019). Design strategies of viscous dampers for seismic protection of building structures: A review, *Soil Dyn Earthquake Eng* (in press).

Garcia, D. L. (2001). A simple method for the design of optimal damper configurations in MDOF structures, *Earthquake Spectra*, 17: 387–398.

Kalkan, E., and Kunnath, S. K. (2006). Effects of fling step and forward directivity on seismic response of buildings, *Earthquake Spectra*, 22(2): 367–390.

Kojima, K., and Takewaki, I. (2015). Critical earthquake response of elastic-plastic structures under near-fault ground motions (Part 1: Fling-step input), *Frontiers in Built Environment*, 1: 12.

Kojima, K., and Takewaki, I. (2016). Closed-form critical earthquake response of elastic-plastic structures with bilinear hysteresis under near-fault ground motions, *J Struct Construction Eng*, AIJ, 726: 1209–1219 (in Japanese).

Kojima, K., Saotome, Y., and Takewaki, I. (2017). Critical earthquake response of a SDOF elastic-perfectly plastic model with viscous damping under double impulse as a substitute of near-fault ground motion, *J. Struct. Construction Eng.*, AIJ, 735, 643-652 (in Japanese). [English Version] *Japan Architectural Review*, Wiley, 2018.

Lavan, O., and Levy, R. (2005). Optimal design of supplemental viscous dampers for irregular shear-frames in the presence of yielding, *Earthquake Eng Struct Dyn*, 34(8): 889–907.

Lavan, O., and Levy, R. (2006). Optimal design of supplemental viscous dampers for linear framed structures, *Earthquake Eng Struct Dyn*, 35: 337–356.

Lavan, O., Cimellaro, G. P., and Reinhorn, A. M. (2008). Noniterative optimization procedure for seismic weakening and damping of inelastic structures, *J Struct Eng*, ASCE, 134(10): 1638–1648.

Mavroeidis, G. P., and Papageorgiou, A. S. (2003). A mathematical representation of near-fault ground motions, *Bull. Seism. Soc. Am.*, 93(3): 1099–1131.

Murakami, Y., Noshi, K., Fujita, K., Tsuji, M., and Takewaki, I. (2013). Simultaneous optimal damper placement using oil, hysteretic and inertial mass dampers, *Earthquakes and Struct*, 5(3): 261–276.

Saotome, Y., Kojima, K., and Takewaki, I. (2018). Earthquake response of 2DOF elastic-perfectly plastic model under multiple impulse as substitute for long-duration earthquake ground motions, *Frontiers in Built Environment*, 4: 81.

Shiomi, T., Fujita, K., Tsuji, M., and Takewaki, I. (2018). Dual hysteretic damper system effective for broader class of earthquake ground motions, *Int. J. Earthquake and Impact Eng*, 2(3): 175–202.

Silvestri, S., and Trombetti, T. (2007). Physical and numerical approaches for the optimal insertion of seismic viscous dampers in shear-type structures, *J. Earthquake Eng*, 11: 787–828.

Singh, M. P., and Moreschi, L. M. (2001). Optimal seismic response control with dampers, *Earthquake Eng Struct Dyn*, 30: 553–572.

Soong, T. T. (1998). "Structural control: Impact on structural research in general", *Proc. of 2nd World Conf. on Structural Control*, 1: 5–14.

Takewaki, I. (1997). Optimal damper placement for minimum transfer functions, *Earthquake Eng Struct Dyn*, 26(11): 1113–1124.

Takewaki, I., Yoshitomi, S., Uetani, K., and Tsuji, M. (1999). Non-monotonic optimal damper placement via steepest direction search, *Earthquake Eng Struct Dyn*, 28(6): 655–670.

Takewaki, I. (2009). *Building Control with Passive Dampers: Optimal Performance-based Design for Earthquakes*, John Wiley & Sons Ltd. (Asia), First Edition.

Taniguchi, R., Kojima, K., and Takewaki, I. (2016). Critical response of 2DOF elastic-plastic building structures under double impulse as substitute of near-fault ground motion, *Frontiers in Built Environment*, 2: 2.

Tsuji, M., and Nakamura, T. (1996). Optimum viscous dampers for stiffness design of shear buildings, *J. Struct Design of Tall Buildings*, 5: 217–234.

Uetani, K., Tsuji, M., and Takewaki, I. (2003). Application of optimum design method to practical building frames with viscous dampers and hysteretic dampers, *Eng Struct*, 25(5): 579–592.

Vamvatsikos, D., and Cornell C. A. (2001). Incremental dynamic analysis, *Earthquake Eng Struct Dyn*, 31(3): 491–514.

Whittle, J. K., Williams, M. S., Karavasilis, T. L., Blakeborough, A. (2012). A comparison of viscous damper placement methods for improving seismic building design, *J. Earthquake Eng*, 16(4): 540–560.

Chapter 12

Future directions

12.1 INTRODUCTION

In this book, an innovative approach was introduced in nonlinear structural dynamics and earthquake-resistant design based on the smart use of impulse and energy balance law. The approach can overcome the long-time difficulty (many repetitions and large computational load) encountered first around 1960 in the field of nonlinear structural dynamics. The critical excitation problems as a worst input for elastic-plastic structures are tackled in a more direct way than the previous methods (Takewaki 2007) using the equivalent linearization methods (Caughey 1960a, b) and the mathematical programming including laborious nonlinear time-history analysis (Moustafa et al. 2010). The approach does not need any time-history response analysis which is believed to be inevitable in nonlinear structural dynamics. The approximate transformation of ground motions into impulses and the smart use of the energy balance approach enabled such dramatic progress. It may be said that the approach opened the door for an innovative field of nonlinear structural dynamics.

In this chapter, some future directions for further development of the approach explained in this book are presented.

12.2 TREATMENT OF NONCRITICAL CASE

In this book, only the critical case (resonant case) was treated. The critical case was characterized by the input timing of the second impulse in the double impulse corresponding to the zero restoring force in SDOF models and the zero restoring force in the first story of the MDOF models. If the energy balance approach is applicable to noncritical cases, it is useful from the viewpoint of avoiding laborious nonlinear time-history response analysis.

In the previous researches, some presentations of such cases were made. Kojima and Takewaki (2015) showed briefly, in their appendix, the expression for noncritical case for demonstrating the criticality of input timing of the second impulse in the double impulse. Kojima and Takewaki (2016a) presented compactly, in their appendix, the expression for noncritical case.

Taniguchi et al. (2017) derived an expression in the nonresonant case for rocking vibration of a rigid block. Homma et al. (2020) investigated the general dynamic collapse criterion for elastic-plastic structures under the double impulse including noncritical case and clarified that the resonant case is not necessarily the critical case giving the lowest limit level of the input velocity.

More advanced developments are desired from the viewpoint of efficient evaluation of the resonance of nonlinear hysteretic vibrating system.

12.3 EXTENSION TO NONLINEAR VISCOUS DAMPER AND HYSTERETIC DAMPER

It is well known that passive dampers, whether viscous or hysteretic, exhibit nonlinear behaviors in the large amplitude range (Soong and Dargush 1997, Hanson and Soong 2001, Lagaros et al. 2013, De Domenico et al. 2019). Tamura et al. (2019) presented the closed-form critical response of elastic-perfectly plastic SDOF systems with nonlinear viscous damping under the multi-impulse as a representative of long-duration ground motions. They used an approximation in the modeling of the damping force of oil dampers in the nonlinear range by extending the method explained in Chapter 5. Shiomi et al. (2016, 2018) demonstrated that the energy balance approach can be applied to the elastic–perfectly plastic SDOF models with single hysteretic and dual hysteretic dampers. The dual hysteretic damper system includes the hysteretic damper for small-amplitude vibration suppression and the hysteretic damper with a gap mechanism for large-amplitude vibration mitigation in parallel. Shiomi et al. (2016, 2018) showed that the present approach can be applied to nonlinear systems with general restoring-force characteristics.

12.4 TREATMENT OF UNCERTAIN FAULT-RUPTURE MODEL AND UNCERTAIN DEEP GROUND PROPERTY

The uncertainties in the modeling of earthquake ground motions are critical for the reliable analysis of seismic safety of structures and infrastructures. In previous studies, many attempts to include various kinds of uncertain parameters have been conducted. However, the viewpoint from the critical response of structures for uncertain parameters in the fault-rupture model and the deep ground model seems missing. To overcome such drawbacks, Makita et al. (2018a, b, 2019) and Kondo and Takewaki (2019) presented some methods for including the uncertainties in the modeling of fault-rupture and deep ground. Application to a broader class of practical structures is desired.

12.5 APPLICATION TO PASSIVE CONTROL SYSTEMS FOR PRACTICAL TALL BUILDINGS

It was demonstrated in the previous works that the present approach is useful for the reliability analysis of advanced passive controlled buildings exploited recently.

The first one is the dual system of base isolation and building connection as shown in Figures 12.1 and 12.2 (Murase et al. 2013, Kasagi et al. 2016, Fukumoto and Takewaki 2017). The base-isolation and the vibration control using passive control devices are two major modern structural systems for building vibration control. It is known that the base-isolated building exhibits a large response to a long-duration, long-period ground motion and the interconnected building without base isolation shows a large response to a pulse-type ground motion as shown in Figure 12.3 (Takewaki et al. 2011, Murase et al. 2013). To compensate for each deficiency, the hybrid passive control system was proposed in which a base-isolated building is connected to another building (free wall/usually car parking) with oil dampers (Murase et al. 2013, Kasagi et al. 2016, Fukumoto and Takewaki 2017).

Fukumoto and Takewaki (2015, 2017) showed that the critical demand of earthquake input energy to connected building structures can be

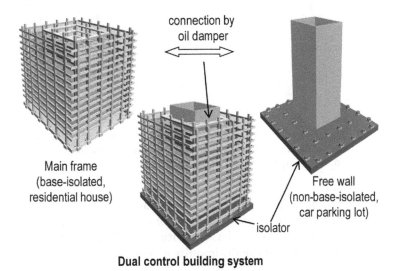

Dual control building system

Figure 12.1 Dual system of base isolation and building connection (Fukumoto and Takewaki 2017).

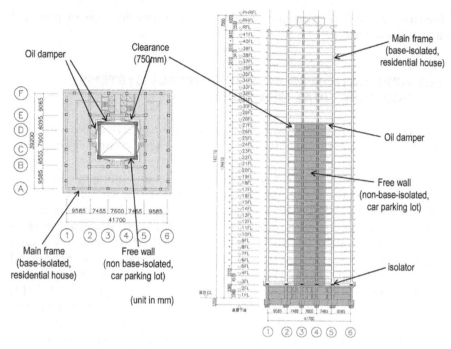

Figure 12.2 Realistic high-rise residential house including base isolation building connection dual control system (Fukumoto and Takewaki 2017).

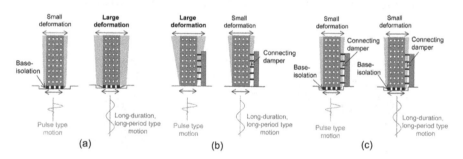

Figure 12.3 Effectiveness of base isolation building connection dual control system: (a) base-isolated building, (b) connected building, (c) base isolation, building connection hybrid system.

evaluated by using the double impulse with variable impulse input timing. Hayashi et al. (2018) demonstrated that the approach is applicable to the simple response evaluation of base isolation building connection hybrid structural systems under the multi-impulse as a representative of long-period and long-duration ground motions.

The second one is the control system for tall buildings using large-stroke viscous dampers through connection to the strong-back core frame usually

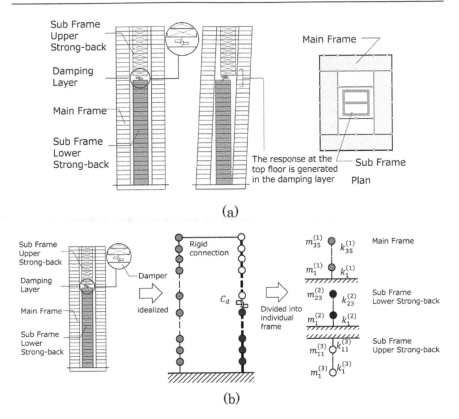

Figure 12.4 Vibration-controlled structure using large-stroke viscous dampers through connection to strong-back core frame: (a) overview of building with control system, (b) simplification of controlled building system into MDOF model and decomposition into sub-assemblage models (Kawai et al. 2020).

designed as car parking, as shown in Figure 12.4 (Kawai et al. 2020). A damping layer is provided between a stiff upper core frame suspended from the top of the main building and a stiff lower core frame connected to the building foundation. As the ratio of stiffness of both core frames to that of the main building becomes larger, the relative displacement in the damping layer (damper deformation) approaches to the top floor displacement of the main building. The large displacement of the top floor of the main building is taken full advantage of in such a control system because most of the total displacement of the main building results from the damper deformation instead of interstory drift. Kawai et al. (2020) presented that the impulse and earthquake energy balance approach is useful for evaluating the critical performance of this control system.

The third one is the control system with shear walls and concentrated dampers in lower stories as shown in Figure 12.5 (Tani et al. 2017).

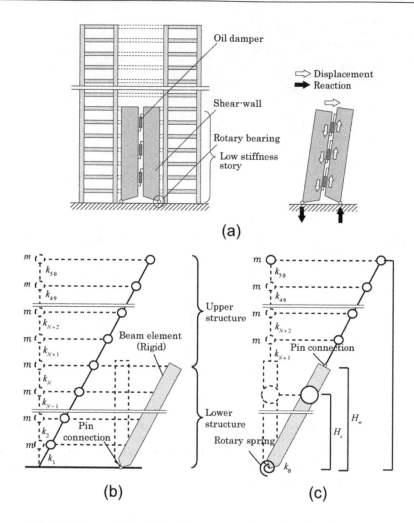

Figure 12.5 Vibration-controlled building structure and its modeling into MDOF model and reduced model: (a) proposed system, (b) MDOF model, (c) reduced model (Tani et al. 2017).

This system has some design variables, i.e., height of shear-wall, degree of main-frame stiffness reduction at lower stories, and quantity of dampers. It is shown that it is possible to prevent from increasing the response under a one-cycle sinusoidal input (impulsive input) resonant to the lowest mode and reduce the steady-state response under a harmonic input (long-duration input) with the resonant fundamental period by reducing the stiffness in the lower structure and increasing the damper deformation. Tani et al. (2017) evaluated the critical performance of this new control system by using resonant sinusoidal waves.

12.6 STOPPER SYSTEM FOR PULSE-TYPE GROUND MOTION OF EXTREMELY LARGE AMPLITUDE

For practical large-scale building structures including above-mentioned advanced structural control systems, it is important to consider unprecedented earthquake ground motions. Pulse-type earthquake ground motions of extremely large amplitude were recorded in the past (for example, Northridge 1994, Kumamoto 2016). While the maximum velocity level of severe earthquake ground motions in the code for tall buildings in Japan is 0.5 m/s, the recorded one in the Kumamoto earthquake (2016) was over 2.0 m/s. These ground motions could cause catastrophic damage to high-rise buildings with a long natural period. To mitigate these damages, some advanced passive damper systems were proposed (Shiomi et al. 2018, Hashizume and Takewaki 2020a, b). Shiomi et al. (2018) proposed the dual hybrid damper (DHD) system consisting of a hysteretic damper (DSA) for the small-amplitude range and another hysteretic damper (DLA) including gap mechanism as a stopper element for the large-amplitude range in parallel. On the other hand, Hashizume and Takewaki (2020a, b) introduced the hysteretic-viscous hybrid (HVH) damper system consisting of a viscous damper and a hysteretic damper (DLA) including gap mechanism as a stopper element in parallel. Shiomi et al. (2018) and Hashizume and Takewaki (2020b) used the double impulse as a representative of pulse-type ground motions of extremely large amplitude and defined the critical double impulse as a set of impulses whose second impulse acts at the time of zero sum of the restoring force and the damper force in the first story in the unloading process. They demonstrated that the DHD and HVH damper systems with a stopper element are excellent for mitigating the catastrophic damages induced by pulse-type ground motions of extremely large amplitude.

Figure 12.6 shows the story shear force–interstory drift relation for the HVH system and the DHD system. "Frame" in the figure indicates the restoring-force characteristic of the main frame. Figure 12.7 illustrates the frame ductility factor distributions of interstory drift of the 30-story shear building model under the critical double impulse of the input velocity levels V=1.0 m/s and 2.0 m/s for various stiffness ratios k_L=0, 1, 2, 4 of DLA to the main

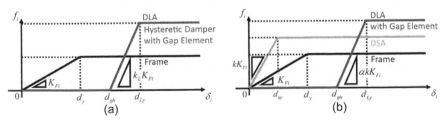

Figure 12.6 Story shear force-interstory drift relation for HVH system and DHD system.

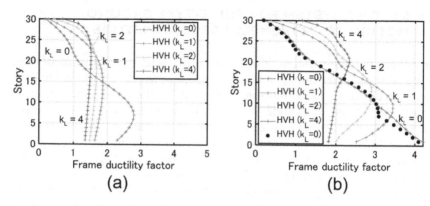

Figure 12.7 Frame ductility factor distribution of interstory drift for various stiffness ratios k_L= 0, 1, 2, 4 of DLA to main frame in 30-story building with HVH system under critical double impulse (solid black circles indicate responses to corresponding one-cycle sine wave) (Hashizume and Takewaki 2020b).

frame whose fundamental natural period is 3.15 s. In this building model, the yield interstory drift d_y=0.0233 m, the fundamental damping ratio of the main frame is 0.02, the fundamental damping ratio of the viscous dampers (frame stiffness-proportional) is 0.05, the uniform floor mass is 4.0×10^5 kg, and the stiffness distribution of the main frame is trapezoidal (top to bottom stiffness ratio = 0.5). It is also noted that DLA starts to work at the ductility factor = 1 ($d_{gb} = d_y$) and yields at the ductility factor=2 ($d_{Ly} = 2d_y$).

It can be observed that, as k_L becomes larger, the frame ductility factor distribution approaches to a uniform one for the input level $V = 1.0$ m/s. In addition, DLA remains in the elastic range for $k_L = 1, 2, 4$. This means that DLA certainly plays a role as a stopper for the pulse-type motion of extremely large amplitude. The solid black circles indicate the responses to the corresponding one-cycle sine wave obtained by the transformation shown in Section 1.2. This good correspondence indicates that the critical double impulse is a good substitute for the corresponding one-cycle sine wave, which is regarded as the main part of pulse-type ground motions.

Further investigation for other types of unprecedented earthquake ground motions will be necessary and the revision of the energy balance law for the MDOF system is expected.

12.7　REPEATED SINGLE IMPULSE IN THE SAME DIRECTION FOR REPETITIVE GROUND MOTION INPUT

The present approach explained in this book can be modified to evaluate the reliability of structures under repeated severe ground motions with extremely

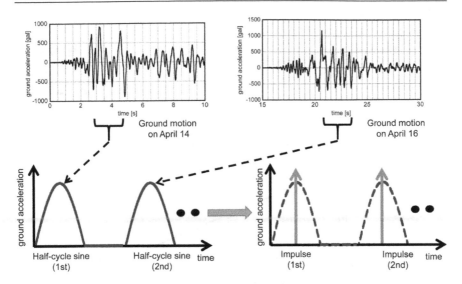

Figure 12.8 Modeling of repeated ground motions into two single impulses (Kojima and Takewaki 2016b).

large amplitude. The repeated input earthquake ground motions with large amplitude are assumed to be modeled by the repetition of single impulse in the same direction as shown in Figure 12.8. Kojima and Takewaki (2016b) showed that the upgrade factor 1.5 for structural strength is necessary for two consecutive ground motions (see Figure 12.9), and Ogawa et al. (2017) demonstrated that larger upgrade factors are necessary for consecutive ground motions with more arrivals. Kojima and Takewaki (2016b) clarified the accuracy and reliability of this upgrade factors for actually recorded ground motions (Kumamoto earthquake in Japan in 2016). The effect of the intensity difference of the repeated ground shakings on the upgrade factors will be the remaining issue together with the influence of the different input directions of the repeated ground shakings.

12.8 ROBUSTNESS EVALUATION

The approach presented in this book can be applied to the robustness analysis (Ben-Haim 2006, Takewaki 2009) of elastic-plastic structures. The most superior measure of robustness may be the robustness function introduced by Ben-Haim (2006), which defines the case satisfying the constraints under the widest uncertainty as the most robust case. Since the robustness analysis for many uncertain parameters and the response analysis of elastic-plastic structures are both time consuming, the approach in this direction has been extremely limited. Kanno and Takewaki (2016) and Kanno et al. (2017) showed that the present approach based on the impulse and the energy

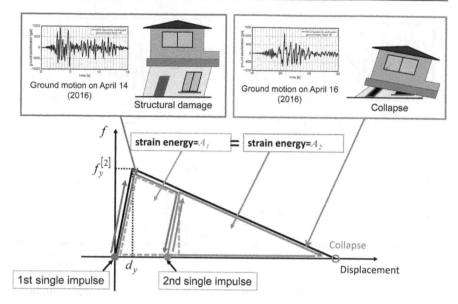

Figure 12.9 Scenario to collapse after repeated ground motions.

balance law is very efficient for the robustness analysis of elastic-plastic structures including some uncertain parameters. Fujita et al. (2017) demonstrated further that the present approach is useful for the robustness evaluation of elastic-plastic base-isolated high-rise buildings under the critical double impulse.

12.9 PRINCIPLES IN SEISMIC RESISTANT DESIGN

In the history of seismic-resistant design, there are two famous hypotheses. One is the constant energy hypothesis and the other is the constant displacement hypothesis (Tanabashi 1935, 1956, Housner 1959, Veletsos and Newmark, 1960, Veletsos et al. 1965, Bozorgnia and Bertero 2004, Riddell 2008, Dowrick 2009, Takewaki 2009, Shibata 2010, Chopra 2012). The former one is applied for buildings with rather shorter natural periods and the latter one is regarded as valid for buildings with longer natural periods. In both hypotheses, the resonance of buildings to input ground motions is not taken into account explicitly. This is because it was believed that earthquake ground motions are random and uncertain.

For buildings with shorter natural periods under random ground motions, they behave as if a short-duration impulsive input is acting on them. In this case, the situation as shown in Figure 1.5 appears. For this reason, the

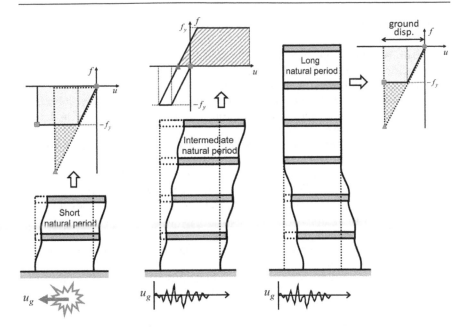

Figure 12.10 Reasoning for three hypotheses (laws), i.e., constant energy law (left), constant displacement law (right), and law for resonant case (middle)

constant energy hypothesis holds approximately regardless of the building response state of elastic or elastic-plastic (Figure 12.10 (left)). For buildings with longer natural periods under random ground motions, an approximation may hold such that the mass (or top location) stays at a fixed point and only ground moves as the earthquake ground motion. In this case, the maximum deformation of the building structure is almost equal to the maximum displacement of the ground. For this reason, the constant displacement hypothesis holds approximately whether elastic or elastic-plastic (Figure 12.10 (right)).

On the other hand, there exists another idea that buildings respond to resonant components of earthquake ground motions and resonance has to be considered appropriately. The earthquake energy balance law used in this book with the help of the introduction of impulses enables the inclusion of such resonance case in the earthquake-resistant design in an explicit manner (Figure 12.10 (middle)).

It is hoped that this novel critical excitation method for elastic-plastic structures plays a role of the third hypothesis for more reliable and robust design of buildings under uncertain earthquake ground motions in the history of earthquake-resistant design.

REFERENCES

Ben-Haim, Y. (2006). *Info-gap decision theory: decisions under severe uncertainty*, 2nd edition, Academic Press, London.

Bozorgnia, Y., and Bertero, V. V. (eds.) (2004). *Earthquake engineering from engineering seismology to performance-based engineering*, CRC Press, Boca Raton, FL.

Caughey, T. K. (1960a). Sinusoidal excitation of a system with bilinear hysteresis, *J. Appl. Mech.*, 27(4): 640–643.

Caughey, T. K. (1960b). Random excitation of a system with bilinear hysteresis, *J. Appl. Mech.*, 27(4): 649–652.

Chopra, A. K. (2012). *Dynamics of structures: Theory and applications to earthquake engineering*, 4th edition, Prentice-Hall, Upper Saddle River, NJ.

Domenico, D. De, Ricciardi, G., and Takewaki, I. (2019). Design strategies of viscous dampers for seismic protection of building structures: A review, *Soil Dynamics and Earthquake Engineering*, 118: 144–165.

Dowrick, D. J. (2009). *Earthquake resistant design and risk reduction*, 2nd ed., Wiley, Chichester, England.

Fujita, K., Yasuda, K., Kanno, Y., and Takewaki, I. (2017). Robustness evaluation of elastic-plastic base-isolated high-rise buildings under critical double impulse, *Frontiers in Built Environment*, 3: 31.

Fukumoto, Y., and Takewaki, I. (2015). Critical demand of earthquake input energy to connected building structures, *Earthquakes and Structures*, 9(6): 1133–1152.

Fukumoto, Y., and Takewaki, I. (2017). Dual control high-rise building for robuster earthquake performance, *Frontiers in Built Environment*, 3: 12.

Hanson, RD and Soong, T. T. (2001). *Seismic design with supplemental energy dissipation devices*, EERI, Oakland, CA.

Hayashi, K., Fujita, K., Tsuji, M., and Takewaki, I. (2018). A simple response evaluation method for base-isolation building-connection hybrid structural system under long-period and long-duration ground motion, *Frontiers in Built Environment*, 4: 2.

Hashizume, S., and Takewaki, I. (2020a). Hysteretic-viscous hybrid damper system for long-period pulse-type earthquake ground motions of large amplitude, *Frontiers in Built Environment*, 6: 62.

Hashizume, S., and Takewaki, I. (2020b). Hysteretic-viscous hybrid damper system with stopper mechanism for tall buildings under earthquake ground motions of extremely large amplitude, *Frontiers in Built Environment*, 6: 583543.

Homma, S., Kojima, K., and Takewaki, I. (2020). General dynamic collapse criterion for elastic-plastic structures under double impulse as substitute of near-fault ground motion, *Frontiers in Built Environment*, 6: 84.

Housner, G. W. (1959). Behavior of structures during earthquakes, *J. Eng. Mech. Div., ASCE*, 85(4): 109–129.

Kanno, Y., and Takewaki, I. (2016). Robustness analysis of elastoplastic structure subjected to double impulse, *J. of Sound and Vibration*, 383: 309–323.

Kanno, Y., Yasuda, K., Fujita, K., and Takewaki, I. (2017). Robustness of SDOF elastoplastic structure subjected to double-impulse input under simultaneous uncertainties of yield deformation and stiffness, *Int. J. Non-Linear Mechanics*, 91: 151–162.

Kasagi, M., Fujita, K., Tsuji, M., and Takewaki, I. (2016). Automatic generation of smart earthquake-resistant building system: Hybrid system of base-isolation and building-connection, *J. of Heliyon*, 2: 2, Article e00069.

Kawai, A., Maeda, T., and Takewaki, I. (2020). Smart seismic control system for high-rise buildings using large-stroke viscous dampers through connection to strong-back core frame, *Frontiers in Built Environment*, 6: 29.

Kojima, K., and Takewaki, I. (2015). Critical earthquake response of elastic-plastic structures under near-fault ground motions (Part 1: Fling-step input), *Frontiers in Built Environment*, 1: 12.

Kojima, K., and Takewaki, I. (2016a). Closed-form critical earthquake response of elastic-plastic structures with bilinear hysteresis under near-fault ground motions, *J. of Structural and Construction Eng.* (AIJ), 81(726): 1209–1219 (in Japanese).

Kojima, K., and Takewaki, I. (2016b). A simple evaluation method of seismic resistance of residential house under two consecutive severe ground motions with intensity 7, *Frontiers in Built Environment*, 2: 15.

Kondo, K., and Takewaki, I. (2019). Simultaneous approach to critical fault rupture slip distribution and optimal damper placement for resilient building design, *Frontiers in Built Environment*, 5: 126.

Lagaros, N., Plevris, V., and Mitropoulou, C. C. (Eds.) (2013). *Design optimization of active and passive structural control systems*, Informaion Science Reference.

Makita, K., Murase, M., Kondo, K., and Takewaki, I. (2018a). Robustness evaluation of base-isolation building-connection hybrid controlled building structures considering uncertainties in deep ground, *Frontiers in Built Environment*, 4: 16.

Makita, K., Kondo, K., and Takewaki, I. (2018b). Critical ground motion for resilient building design considering uncertainty of fault rupture slip, *Frontiers in Built Environment*, 4: 64.

Makita, K., Kondo, K., and Takewaki, I. (2019). Finite difference method-based critical ground motion and robustness evaluation for long-period building structures under uncertainty in fault rupture, *Frontiers in Built Environment*, 5: 2.

Moustafa, A., Ueno, K., and Takewaki, I. (2010). Critical earthquake loads for SDOF inelastic structures considering evolution of seismic waves, *Earthquakes and Structures*; 1(2): 147–162.

Murase, M., Tsuji, M., and Takewaki, I. (2013). Smart passive control of buildings with higher redundancy and robustness using base-isolation and inter-connection, *Earthquakes and Structures*, 4(6): 649–670.

Ogawa, Y., Kojima, K., and Takewaki, I. (2017). General evaluation method of seismic resistance of residential house under multiple consecutive severe ground motions with high intensity, *Int. J. Earthquake and Impact Engineering*, 2(2): 158–174.

Riddell, R. (2008). Inelastic response spectrum: Early history, *Earthquake Engineering and Structural Dynamics*, 37: 1175–1183.

Shibata, A. (2010). *Dynamic analysis of earthquake resistant structures*, Tohoku University Press (Japanese version in 1981).

Shiomi, T., Fujita, K., Tsuji, M., and Takewaki, I. (2016). Explicit optimal hysteretic damper design in elastic-plastic structure under double impulse as representative of near-fault ground motion, *Int. J. Earthquake and Impact Engineering*, 1(1/2): 5–19.

Shiomi, T., Fujita, K., Tsuji, M., and Takewaki, I. (2018). Dual hysteretic damper system effective for broader class of earthquake ground motions, *Int. J. Earthquake and Impact Engineering*, 2(3): 175–202.

Soong, T. T., and Dargush, G. F. (1997). *Passive energy dissipation systems in structural engineering*, JohnWiley & Sons, Chichester.

Takewaki, I. (2007). *Critical excitation methods in earthquake engineering*, Elsevier, Second edition in 2013.

Takewaki, I. (2009). *Building control with passive dampers: Optimal performance-based design for earthquakes*, John Wiley & Sons Ltd., Asia.

Takewaki, I., Murakami, S., Fujita, K., Yoshitomi, S., and Tsuji, M. (2011). The 2011 off the Pacific coast of Tohoku earthquake and response of high-rise buildings under long-period ground motions, *Soil Dynamics and Earthquake Engineering*, 31(11): 1511–1528.

Tamura, G., Kojima, K., and Takewaki, I. (2019). Critical response of elastic-plastic SDOF systems with nonlinear viscous damping under simulated earthquake ground motions, *Heliyon*, 5: e01221.

Tanabashi, R. (1935). Personal view on destructiveness of earthquake ground motions and building seismic resistance, *J. Archit. Build. Sci.*, 48: 599 (in Japanese).

Tanabashi, R. (1956). *Studies on the nonlinear vibrations of structures subjected to destructive earthquakes*, Proc. of the First World Conf. on Earthquake Engineering, Berkeley, CA, 6:1–6:16.

Tani, T., Maseki, R., and Takewaki, I. (2017). Innovative seismic response controlled system with shear wall and concentrated dampers in lower stories, *Frontiers in Built Environment*, 3: 57.

Taniguchi, R., Nabeshima, K., Kojima, K., and Takewaki, I. (2017). Closed-form rocking vibration of rigid block under critical and non-critical double impulse, *Int. J. Earthquake and Impact Engineering*, 2(1): 32–45.

Veletsos, A. S., and Newmark, N. M. (1960). Effect of inelastic behavior on the response of simple systems to earthquake motion, *Proc. of the Second World Conference on Earthquake Engineering*, Science Council of Japan, Japan, 2, 895–912.

Veletsos, A. S., Newmark, N. M., and Chelapati, C. V. (1965). Deformation spectra for elastic and elasto-plastic systems subjected to ground shock and earthquake motions, *Proc. of the Third World Conference on Earthquake Engineering*, New Zealand, 2, 11.663–11.682.

Index

9 780367 681418